THE FACT OF EVOLUTION

THE FACT OF EVOLUTION

CAMERON M. SMITH

59 John Glenn Drive
Amherst, New York 14228–2119

Published 2011 by Prometheus Books

Cover design by Nicole Sommer-Lecht

Inquiries should be addressed to
Prometheus Books
59 John Glenn Drive
Amherst, New York 14228–2119
VOICE: 716–691–0133
FAX: 716–691–0137
WWW.PROMETHEUSBOOKS.COM

15 14 13 12 11 5 4 3 2 1

Library of Congress Cataloging-in-Publication Data

Smith, Cameron McPherson, 1967–
The fact of evolution / by Cameron McPherson Smith.
p. cm.
Includes bibliographical references and index.
ISBN 978–1–61614–441–8 (alk. paper)
ISBN 978–1–61614–442–5 (e-book)
1. Evolution (Biology) I. Title.

QH367.S58 2011
576.8—dc22

2010049704

Printed in the United States of America on acid-free paper

DEDICATION AND ACKNOWLEDGMENTS

I thank, and dedicate the energies expended in this work to, the legions of biologists, geneticists, naturalists, and field researchers who have worked for over a century in relative obscurity to help humanity better understand the world of living things.

I also thank my parents, professors Donald E. and Margit J. Posluschny Smith, for building a calm and stable world for their children; and my brothers, Mark J. and D. Julian Smith, for taking an interest in my work and making family life exciting. Finally, I am thrilled to thank Angela Perri for giving me the time and energy required to write this book.

CONTENTS

CONTENTS

LIST OF FIGURES

LIST OF TABLES

FOREWORD

I first started thinking about evolution in 1987, when I was an undergraduate student at Harvard University's Koobi Fora Field School near Lake Turkana, Kenya. Crawling across the desert in search of early hominin fossils, I wasn't always as attentive as I might have been to the Great Hunt. I worried about cobras, for instance—one had nearly bitten Dr. Meave Leakey in the outhouse—and occasionally I scanned the terrain for them, taking my eyes off the all-important ground.

When we found million-year-old stone tools, I wondered about the life of the hominin that had chipped and used those tools so long ago. What did those hominins worry about? What did they have to think about if they were going to survive? I was learning that African landscapes weren't simply a stage on which early hominins acted out their lives; instead, hominins were members of a complex ecology. We could understand some part of hominin evolution from the fossils of early hominin bones themselves, but we also had to consider their place in an entire ecosystem. Back in England, I expanded my studies in order to understand something about those ecosystems, and I read a fascinating book by Robert Foley, *Another Unique Species*, which really put the hominins into a larger context. Things were getting very complicated. And then I read another book, Richard Dawkins's *The Selfish Gene*, and that opened a door to another dimension entirely: that of the molecular world of the gene.

In the late 1980s, the bulk of early hominin studies was the study of fossils, but soon new techniques made the study of human evolutionary genetics a reality. And things became yet more complex—and fascinating—as through my graduate studies in anthropology and archaeology human evolution was examined on the scales of the molecule, the individual hominin, populations of hominins, and those populations as members of ecosystems.

While my studies diverged into the archaeology of the Pacific Northwest Coast, I always kept up with developments in evolutionary biology. I had to; evolution was the explanatory device for all life, "even" humanity, which is, in many ways, so unique.

As I began teaching courses on human evolution and prehistory in general, I was alarmed to find that many times my students held fundamental misconceptions about evolution. It was (and remains) rewarding to correct some of these misconceptions and give my students a better understanding of the evolutionary process that has shaped not only their own bodies but the world they live in—and to better understand it myself, because teaching is as much about learning as it is about imparting knowledge and the skills of critical thinking.

But while everyone is the product of evolution, not everyone has the opportunity to study evolution, and so to reach out beyond my students and dismantle these misconceptions for a wider audience I wrote a popular-science book with a fellow educator, Charles Sullivan, titled *The Top Ten Myths about Evolution*. I am thrilled to find that our little book continues to circulate worldwide, being now held by nearly a thousand libraries from Arizona to New Zealand and endorsed by the American Library Association, the American Association for the Advancement of Science, the National Center for Science Education, and several well-respected popular-science writers including Ann Druyan, who coauthored the *Cosmos* television series and several books with the late astronomer and popularizer of science, Carl Sagan.

Foreword

I felt passionately that I had to write *The Top Ten Myths about Evolution*. Common misconceptions drove people to reject evolution and even the scientific approach that went with it. In a civilization built by science—from cell phones to gene therapy to space shuttles—that disjunction was disturbing. And as separate as we humans might feel from Nature (with a capital N), we do not exist apart from it; every mile we drive is at the courtesy of Nature, and every calorie we eat—every calorie that allows us to get to an airport or just across the street—comes from Nature. I don't want everyone to be a scientist, but I do want to try to improve the understanding of evolution because—at the least—understanding of Nature has real-world consequences when we consider biodiversity, gene therapy, the very food we eat, and everything else related to living things.

The Top Ten Myths about Evolution is about what evolution *isn't*. This book, *The Fact of Evolution*, is about what evolution *is*.

Now, *What Evolution Is* happens to be the title of biologist Ernst Mayr's great 2005 introduction to evolution, so one might reasonably ask whether another such book is needed; I certainly had to justify it to my publisher. I feel strongly that it is, and I am happy that Prometheus Books felt the same way. Despite many excellent books, like Mayr's, evolution continues to be rejected by many, often, I think, because of misunderstanding. I think education can overturn misunderstanding—I have seen it in my own classes—so I continue on with this book.

I picked the title of this book, *The Fact of Evolution*, to make it clear that the "evolution debate" is over. Evolution isn't "just a theory" but a fact accepted by widespread consensus in the life sciences. Evolution is, and should be discussed as, a fact. And explaining why it is accepted as a fact is important; it's not a fact because some high academic declared it a fact, it is a fact because thousands of studies over 150 years since Darwin have confirmed it as a fact with innumerable kinds of evidence.

Also, I often feel that evolution isn't as clearly presented as it could be. Introductory books don't usually follow the linear, clearly organized approach I've developed for this book. My method of presenting evolution has been developed for my lectures over the past ten years, and I've had great success in making the essence of evolution absolutely clear to many students. I hope this book does the same for its readers.

Thomas Bulfinch (1796–1867) opened his 1855 book, *Bulfinch's Greek and Roman Mythology: The Age of Fable*, with the following words:

> Without a knowledge of mythology much of the elegant literature of our own language cannot be understood and appreciated. When Byron calls Rome "the Niobe of nations," or says of Venice, "She looks a Sea-Cybele fresh from the ocean," he calls up to the mind of one familiar with our subject, illustrations more vivid and striking than the pencil could furnish, but which are lost to the reader ignorant of mythology. . . . This is one reason why we often hear persons by no means illiterate who say they cannot enjoy Milton. . . .[1]

I feel the same applies to understanding the natural world of living things; appreciation of evolution gives us a richer understanding of every living thing, and we are living things surrounded by living things. How could we not be interested in how we come to be or how our food comes to be? How do we get to become so interested in the world of living things? It might take simply opening a book:

> The page he opened was under the heading of Anatomy, and the first passage that drew his eyes was on the valves of the heart. He was not much acquainted with valves of any sort, but he knew that *valvae* were folding-doors, and through this crevice came a sudden light startling him with his first vivid notion of finely-adjusted mechanism in the human frame. A liberal education had of course left him free to read the

indecent passages in the school classics, but beyond a general secrecy and obscenity in connection with his internal structure, had left his imagination quite unbiased, so that for anything he knew his brains lay in small bags at his temples, and he had no more thought of representing to himself how his blood circulated than how paper served instead of gold. But the moment of vocation had come, and before he got down from his chair, the world was made new to him by a presentiment of endless processes filling the vast spaces planked out of his sight by that wordy ignorance which he had supposed to be knowledge. From that hour Lydgate felt the growth of an intellectual passion.[2]

PREFACE

Sir Mortimer Wheeler potently quipped that "[c]hronology is the backbone of archaeology; not the entire skeleton, but nothing less than the backbone." In this book, I am presenting the backbone of evolution; not the whole skeleton—not even the flesh—but nothing less than the backbone. As a popular-science writer, I find the specter of oversimplification is always at my heels. Painters find a place between abstraction and realism; professors and popular-science writers must do the same.

Many of the endnotes in this book are elaborations on a particular theme, some provide reports from which I draw my examples, and most point the reader to further readings. Some clarify or give context. Some endnotes have the character of a box of chocolates: you never know what you're going to get.

In three years of research for this book, I have been thrilled and overwhelmed by the diversity of research on evolutionary topics in every domain in the great world of living things. I've been careful to provide examples in this book from as many as I can, but many kinds of life are underrepresented. Most obviously absent are the microbes and the plants, though I do mention them. I've carefully avoided, though, using examples from human evolution. Humans have, of course, evolved, but since we began to rely on tools and culture to survive over two million years ago, many of the details of our evolution have been significantly different from those of other life-forms. We multiply beyond our immediate resources, we have extremely complex rules regarding reproduction, and we shield ourselves from many rigors of the natural world—factors I'll later call "selective pressures"—by inventing artifacts and methods with a consciousness that appears to be absent from many other life-forms. Indeed, these facts have, I argue in chapter 9, preconditioned us humans to find evolution hard to understand. For all these reasons, I prefer to look at evolution—stripped down to the consequences of replication, variation, and selection—from the perspective of the many millions of other forms of life with which we share our planet.

I've specifically avoided discussing human genetics; so much research of the human genome is related to identifying the causes of disorders and other problems (for example, cancer) that reading about human genes one gets the impression, as author and journalist Matt Ridley has noted, that the genes are there simply to cause problems for people. Instead, I discuss the genetics of many nonhuman life-forms.

I also tend to avoid using the term *species*, although I do define it when necessary. Instead I normally use the term *life-form*, sometimes to refer to an individual, sometimes to refer to a whole group or "kind" of life (a species).

INTRODUCTION

There is no such thing as evolution.

I write this not to be perverse (considering the title of this book) or cute or obtuse, but to make a very important point that I think will help you better understand the natural world.

When we hear the word *evolution* on TV or read about it in the mass media, it's hard to avoid the impression that evolution is a thing. Not quite a noun—it doesn't have the concreteness of a person or a place—but evolution comes across as a process with a kind of "thingness." Evolution, we're told, "sculpts" species to perfectly fit their environments. Evolution, we're told, "produced" the human race. Evolution, we're told, "only cares about" the survival of the fittest. There's an implied "itness" woven deeply into these descriptions of evolution. But this "itness" is a mirage.

This "itness" even implies an *intent* that, as we'll see throughout this book, is simply absent. Even in such a wonderful book as Matt Ridley's *Genome*, we read that some early human ancestors "lost their [large] canine teeth the better to chew from side to side."[1] But, as we'll see, these early humans didn't lose their larger canine teeth *to better do* anything; all that happened is that those early human ancestors that were born—at first just by chance—without the larger canines of their ancestors, flourished, for whatever reasons, sending the genes for somewhat smaller canines on into the future, eventually "swamping" the genes for larger canines. There was no evolution there, deciding that it would be best to do this or that with the canine teeth. The wording here is very subtle, but its power—as in poetry—lies exactly in that subtlety.

My perspective is that however true these statements may be—yes, evolution is the explanation for the shapes and diversity of life-forms—the way they're worded often fuels a fundamental misunderstanding of evolution. To clarify things, in this book I explain the principles of evolution in a way that takes the word apart, allowing us to see inside the "black box" of *evolution*.

Evolution is a word we use to characterize the unintended consequences of three independent facts of the natural world:

1. The fact that life-forms have offspring (replication)
2. The fact that offspring are not identical (variation)
3. The fact that some offspring pass more of their genes on to the next generation than do other offspring (selection)

Evolution is not a thing; it is the *consequence* of these three observable facts of *replication*, *variation*, and *selection*. And because evolution is the logical consequence of these three facts, it is no longer debatable; evolution doesn't just happen, it *has to* happen.

This book explains how we know this to be true.

CHAPTER 1
NULLIUS IN VERBA

Evolution is change, adaptation to new circumstances. Evolutionists see a world of process, of flux, incomplete, imperfectly known.

—A. Kehoe, *What Darwin Began: Modern Darwinian and Non-Darwinian Perspectives on Evolution*[1]

In January 2008, the world's leading scientific journal, *Nature*, carried an editorial titled "Spread the Word: Evolution Is a Scientific Fact, and Every Organization Whose Research Depends on It Should Explain Why."[2] Speaking for the world of science, *Nature* publicly established evolution as a fact, in black and white, and called on educators worldwide to teach it as a fact.

WHY DID IT TAKE SO LONG?

Evolution has been referred to as a fact by many, and for a long time. Today evolution is considered a given in the life sciences, as gravitation is considered a given in physics. A universe of detail remains to be investigated in the domain of evolution, and discoveries in the life sciences, which occur every single day, ensure that our understanding of evolution is constantly updated. Even twenty-five years ago evolutionary biologist Niles Eldredge wrote:

> Evolutionary theory is currently in a state of flux. There is far less agreement on basic elements of evolutionary theory now than there was ten years ago. . . . Though some biologists may long for the halcyon days when nearly everyone agreed on the essentials of a single, simple and quite elegant evolutionary theory, the zest for renewed explorations in evolutionary theory is more than adequate compensation. The fervor of argument in evolutionary biology these days is the surest sign of its intellectual health: evolutionary theory, perhaps now more than ever before, is an active, vital, and truly scientific endeavor.[3]

Despite occasional mass-media hype that some new finding means that "evolution is dead," evolution is certainly not dead; it is actually buzzing with life, as we'll see throughout this book. Still, drama sells; a recent *National Geographic* issue carried a cover story titled "Was Darwin Wrong?" The article concluded that he was not wrong about the basic principles of evolution, but the striking title is memorable.[4]

Charles Darwin (1804–1882), who proposed the basics of evolution as we understand them today, might well have been wrong about many things; indeed, he lived before the genes, as we know them today, were even understood. But, as we'll see, he was not wrong about the basic facts of the evolutionary process. For over 150 years, scientists have scrutinized Darwin's theory, searching the natural world—from the seafloor to the high mountains—for anything that would disprove it, and they have tested it with laboratory experiment. While many other theories have been rejected in the past 150 years, Darwin's has not.

If things are so clear, why did science—spoken for at large by *Nature*—take so long to publicly declare evolution a fact? There are at least two reasons.

SCIENCE IS SLOW

While there are many scientists and they discover new things every day, science at large—meaning general consensus—moves slowly. A century or a century and a half is not so long; the great biologist Ernst Mayr (1904–2006) was born only twenty-two years after Darwin's death, and modern field studies consume entire careers. Another titan of biology, Carl Woese (b. 1928), who worked for years on the problem of the basic classification of life-forms, recently described his years of labor:

> You got up in the morning, ate breakfast and came into the lab. And then you put these silly X-rays up on your wall. And you looked intensely. But you had to be intense to get through this task. This happens on Monday, Tuesday, Wednesday, Thursday, Friday, week after week and ultimately year after year. I would finish the day, having looked at these films for many many hours, and I would go home saying to myself "You have destroyed your mind again today." That's how it felt. . . . All my mental energy was used up.[5]

This went on for ten years, but it culminated in a fascinating new understanding of the origins and evolution of life. The scientific community—who demand evidence for claims (as characterized by the Royal Society's motto, *Nullius in Verba*, loosely translated as "take nobody's word for it"[6])—are normally slow to overturn basic concepts.

AS GOES AMERICA, SO GOES THE WORLD

Nature is a British journal, and in Britain there has been far less public acrimony over the "theory of evolution" than here in the United States. Why not, then, declare evolution a fact earlier? Because in Britain the very lack of debate caused evolutionary principles to be accepted as fact early on.[7] The obvious did not need stating. While the details of Darwinian evolution have been debated from day one, Darwin's essential points were so convincing that they were rapidly accepted in Britain and on the continent. Darwin published *On the Origin of Species* in November 1859, and by 1863 biologist T. H. Huxley (1825–1895) wrote that "all other theories are absolutely out of court."[8] And these other theories (we'll come to the distinction between fact and theory below) were not "out of court" simply because Huxley greatly admired Darwin. They were "out of court" because year by year alternatives to Darwinian evolution, based on worldwide, independent studies focusing on independent lines of evidence—such as *embryology* (the development of life-forms before birth), *paleontology* (the study of ancient life-forms, performed in the 1800s by examining fossils), and *botany* (the study of plants)—all pointed (as Huxley noted and as we'll see in this book) *in the same direction*; Darwin was right.

Things were very different in the United States, where Darwin's ideas took more than a decade to catch on widely. But by the 1920s, when most American scientists believed that Darwin was essentially correct, a strong opposition to evolution appeared. It is interesting (and transparently self-serving) that this opposition did not come from any branch of science; it was not ecologists (who study ecosystems) or ornithologists (the studiers of birds), for example, who opposed evolution with stacks of new data. Instead, the opposition to the science of evolution came from a completely different domain, that of religion.

Religion does not normally reach into, say, the domains of plumbing or aircraft engineering or songwriting; those are outside its practical knowledge. How, then, could it claim to have anything to say about biology? The answer is very old and serves even today; for the religious fundamentalist (of whatever faith), humanity is a special creation, a product of the divine mind of God. But Darwinian evolution implies that humanity is actually one of many millions of life-forms that have appeared on Earth in the last few billion years. The two approaches were for a long time essentially irreconcilable, but as early as the 1940s the evangelical American Scientific Association reversed its early rejection of evolution, negotiating that evolution occurred but was occasionally mediated by divine intervention.[9] This is a position that at least allows some scientists and some evangelicals to co-exist. But since that time a vocal and often well-funded religiously based opposition to evolution has flourished in the United States. Remember, these groups do not provide new data or logical disproofs of evolution. They always, ultimately, argue that their bible is inerrant and that evolution, not conforming to the Bible, must be in error.[10] Such critiques continue today. What do they have to do with Britain's journal *Nature* declaring evolution a fact in 2008?

Ultimately it has to do with global interconnection. The Internet and satellite communications have "shrunk" the world. What people do in one country has strong and immediate effects in other countries. With the rise of American antievolutionism in the last decade, it was *Nature*'s responsibility to declare evolution a fact in the public sphere. After giving the US National Academy of Sciences "three cheers" for publishing a position paper calling evolution a fact,[11] the *Nature* editorial continued: "Creationism is strong in the United States and, according to the Parliamentary Assembly of the Council of Europe, worryingly on the rise in Europe."[12]

As goes America, so goes the world (at least in some things). Antievolutionism is often linked with antiscience itself, which would have the human mind retreat to an essentially medieval worldview. Enough was enough; *Nature* had to establish evolution as a fact in public.

THE FACTUAL STATUS OF EVOLUTION

I mentioned before that the Royal Society's motto is *Nullius in Verba*, or (essentially) "take no person's word for it," indicating science's demand for data to back up claims. Extraordinary claims, the late astronomer Carl Sagan liked to say, demand extraordinary evidence. This demand is one of the hallmarks of scientific thought, rooted in Renaissance principles of free inquiry, and it relates generally to the scientific method of generating knowledge. Science recognizes no real authority. No matter how erudite or well educated you are, no matter whom you've studied with or whom you know, if your data don't back up your ideas, nobody is going to believe you. *Nullius in Verba*. Show me the beef, we said in the 1980s; show me the money, in the 1990s. *Nullius in Verba* for over four hundred years. In *Paradise Lost* by John Milton (1608–1674), the archangel Michael says to Adam: "Be lowly wise: Think only what concerns thee and thy being."[13] A greater contrast with science could not be found.

So, just because *Nature*—or any other respected journal—declares evolution a fact does not *make* it a fact (if you could make things factual simply by declaring them, our universe would be strange indeed). *Nature*'s declaration tells us only that there is consensus among the scientific community that Darwinian evolution does occur; that it is a fact.

Why the consensus? Easy. As even T. H. Huxley noted before 1900, so much evidence points in the same direction; Darwin was right. We'll look more closely at how science (and evolutionary science) works, but for the moment, keep in mind that two aspects of scientific knowledge-generation make testing out ideas—at least in principle—straightforward. These are *testing to disprove* an idea, and *independent verification* of ideas. We'll see each below in a brief look at how science generates knowledge, including facts.

HYPOTHESES, THEORIES, AND FACTS

In science, if you observe something, like the fall of a pen to the floor, you may devise a *hypothesis* to explain it. A hypothesis is usually a statement of your belief about the relationship of something to something else. For example, my hypothesis about the pen falling to the floor may be that "pens are attracted to carpets." That sounds ridiculous because we have plenty of evidence to the contrary; and in the same way, "nutty" hypotheses can often be quickly spotted and weeded out by what science already knows. But let's say you don't know anything about the properties of pens and carpets, and this hypothesis is considered reasonable. It hasn't been shown to be true; it's just passed your initial "baloney detector," and you decide to move on to the next method of testing hypotheses.

Each hypothesis—remember, just a statement about the relationship of something to something else, normally—can have *test implications*. These are things you would expect to observe if your hypothesis is correct and things you would expect to observe if your hypothesis is not correct. If you do a number of tests releasing pens above a carpet, and the pens repeatedly land on the carpet, you haven't learned much. If all you needed to prove something were a demonstration of what you *already* believe, well, again, think of how strange a universe it would be. No, for science you need to refine your hypothesis for a test implication that would show that your hypothesis is *wrong*; a test to *disprove*. For example, I could come up with a test implication to *disprove* my own hypothesis about pens being attracted to carpets. If you were to stand in a wood-floored room in which some carpet has been mounted on a wall, and then hold the pen next to the carpet on the wall, and then release the pen, if the pen hits the floor rather than the carpet, you can say that your hypothesis about pens being attracted to carpets is suspect. You could continue to disprove the hypothesis by mounting carpet on the ceiling and letting go of the pen, and so on. Eventually you would reject the hypothesis that pens are attracted to carpets because they do not always move toward carpets. Instead, you rework your hypothesis to state that "all other things being equal, less massive objects (like the pen) are attracted to more massive objects (like the earth)." Now you devise test implications to *disprove* that hypothesis, and you run those tests. But they don't disprove it; time and again, no matter how hard you try, you can't disprove the hypothesis. Over time you may become so certain about its explanatory value that you take it for granted in your other investigations. It can always be questioned and tested, but you've done that, and it is reasonable to move on.

Another way to evaluate an idea or hypothesis is to have it tested independently. This means having people other than yourself do the tests, to guard against bias; it also means devising *other* ways to test the hypothesis than were originally used. If testing to disprove doesn't disprove, and other scientists are coming up with the same results—even when they devise the most diabolically clever ways to disprove an idea—scientists begin to accept that they have discovered a reality. They move toward considering the hypothesis or idea *confirmed*.

Confirmed hypotheses are considered knowledge, and that knowledge is used in the generation of larger, more overarching explanations of observations, which we call *theories*. So, for example, many tests show that your less-massive to more-massive attraction hypothesis is so useful that you might combine it with other hypotheses to form a larger explanatory device called *gravitational theory*, which explains a great many observations; in fact, it's so accurate a description of the relationships between things that it can be used to make very precise predictions about, say, launching an airplane or using the force of gravity to "slingshot" space probes throughout the solar system.

In the same way, *evolutionary theory* is so named because it is a proposed explanation of many, many observations. And it is a fact because so many tests that might disprove it have been done, and it has not been disproven; we'll see that throughout this book.

As important as words are, a lot of the wording here does not matter.

What matters is, first, whether or not there is a world external to humanity; and the word from science is, yes, there is. The planet Saturn—which existed independent of humanity long before humanity existed—would not cease to exist if humanity were to become extinct. We humans did not invent Saturn; we discovered it. Second, if there is a reality external to humanity, can we learn about it and make generalizations about how that reality works? Again, the answer is yes; aircraft do not fly because of some supernatural power, they fly because we have learned to build wings that provide lift in certain circumstances. Whether we call our learnings hypotheses, theories, facts, or laws is not as important as knowing that we can indeed learn about the universe we live in, including the universe of living things.

In the case of evolutionary theory—which generally states that the properties of life-forms change through time—evidence for it (and none overturning it) has been accumulated from all branches of the life sciences, from *ichthyology* (the study of fishes) to *mycology* (the study of fungi), and they all point in the same direction. And new studies, almost undreamed-of in Darwin's time, completely independently also point in the same direction; the wholly new field of *genomics*, the study of the molecular properties of different life-forms, shows again and again, thousands of times each day in laboratories worldwide, that Darwin was right. The molecular biologist studying gene sequences in New York, for example, has no connection—none whatsoever—with an Australian biologist tracking kangaroo mating behavior; however, their findings both point in the same direction; Darwin was right. And this molecular data is compelling evidence indeed, because it directly corroborates what all the other data sources are saying; Darwin was right. If Darwin were wrong in his very simple proposal, it would be immediately obvious to a thousand researchers.

Finally—and this is not at all trivial—remember that scientists are human beings with all manner of human motivations, including in some cases a craving for status and recognition. If an easy toppling of Darwinian evolution were possible (even if it would take a whole career), you can bet that scientists—from any scientific field at all—would be scrambling for the Nobel Prize and international fame and fortune that would inevitably be theirs. Why would the Nobel be awarded for toppling evolution? Because science recognizes that it is fallible, that it is subject to revision. A stronger contrast with religiously generated knowledge is not possible. Scientific heretics are praised if ultimately proven correct because they show us new truths. In the strongest possible contrast, religious heretics are punished because they stray from what is believed to be the inerrant and complete truth, said to reside in their ancient and unchanging texts.

The evidence for evolution, then, has been generated over a long period and subjected to intense and careful scrutiny. It comes from sources as different as fossils excavated in East Africa and fly genes observed in a lab in Michigan. Also, we can see the processes of evolution occurring every day in the natural world. Finally, nobody has toppled evolutionary theory despite its essential simplicity. These are the reasons science refers to evolution as a fact.

It is good to see that despite the influence of religious critics, in the United States the factual status of evolution has also recently been widely published. The American Association for the Advancement of Science states: "The theory of biological evolution is more than 'just a theory.' It is as factual an explanation of the universe as the atomic theory of matter or the germ theory of disease. Our understanding of gravity is still a work in progress. But the phenomenon of gravity, like evolution, is an accepted fact."[14]

And in the free booklet *Science, Evolution, and Creationism*, the US National Academy of Sciences writes: "This booklet is also directed to the broader audience of high-quality school and college students as well as adults who wish to become more familiar with the many strands of evidence supporting evolution and to understand why evolution is both a fact and a process that accounts for the diversity of life on Earth."[15]

WHY IS EVOLUTION STILL DISBELIEVED OR MISUNDERSTOOD?

With evolution clearly the consensus belief as the organizing principle for all biology, how is it, then, that only about half of Americans believe it occurs? The answers are simple.

First, some very powerful religious groups continue to reject evolution. These critiques are so transparent and self-serving because they attack only one position in science—evolution, which relates to the origin of humanity. But exactly the same system of observation, logic, inference, and experimental generation of knowledge that leads us to evolution leads us to the engineering of aircraft, computers, medicines, and every single one of the other science-based foundations of our civilization. Ultimately, religious critics wish to maintain human specialness, a central place for humanity in the universe. But astronomy shows we are not at the center of the universe or even our solar system. We orbit the sun, which is one of billions of stars. And biology shows that while humanity certainly is unique in many ways, humanity is not the end of evolution, its pinnacle, or something that all life-forms are "trying" to evolve into. Every life-form follows its own path; we are one of millions.

Second, evolution is still widely referred to as a "theory," and while it is a theory—an explanation of observed facts—it is indeed also a fact. It occurs, as we'll see throughout this book. But as long as we refer to it as a theory, it will remain misunderstood because of the common misunderstanding of the word *theory* as an unsubstantiated guess. Hence the *Nature* editorial with which we began this chapter, and hence the title of this book.

Third (I will argue this point later), part of human identity, part of what makes us, us, are invention and creation. For over two million years, humanity has survived by making things, by inventing solutions to problems. The idea that the natural world could arise only from similar creations, and not natural processes, is culturally and linguistically ingrained.

In a recent interview with *Discover* magazine, evolutionary biologist Sean Carroll summed things up this way:

> It is a cultural issue, not a scientific one. On the science side our confidence grows yearly because we see independent lines of evidence converge. What we've learned from the fossil record is confirmed by the DNA record and confirmed again by embryology. But people have been raised to disbelieve evolution and to hold other ideas more precious than this knowledge. At the same time, we routinely rely on DNA to convict and exonerate criminals. We rely on DNA science for things like paternity. We rely on DNA science in the clinic to weigh our disease risks or maybe even to look at prognoses for things like cancer. DNA science surrounds us, but in this one realm we seem unwilling to accept its facts. Juries are willing to put people to death based upon the variations in DNA, but they're not willing to understand the mechanism that creates that variation and shapes what makes humans different from other things. It's a blindness. I think this is a phase that we'll eventually get through. Other countries have come

to peace with DNA. I don't know how many decades or centuries it's going to take us.[16]

There is quite a gulf between minds that consider humanity central to the universe and designed for a purpose, and minds that consider humanity to be the product of the natural process of evolution. Some of the differences between these conceptions of the universe and our place in it are provided in Table 1-1.

Table 1-1. Ancient and Modern Conceptions of the Universe

Domains or Concepts	Ancient	Modern
Nature of Universe	Static/Unchanging	Dynamic/Changing
Place of Humanity in Nature	Central	One of Many
Structure of Life-Forms	Functional/Designed for Specific Roles	Unintended Consequence of Natural Processes
Age of Universe and Earth	Recent (Shallow Time)	Ancient (Deep Time)
Essence of Life	Vital/Supernatural	Elemental/Natural
Humanity's Knowledge	Scripture & common sense make for complete knowledge: "Be Lowly Wise"	Incomplete; common sense is distrusted; Demand for evidence; "*Nullius in Verba*"
Knowledge Generation	Re-analysis of unchanging scripture	Observations & experiments lead to constant accumulation of new knowledge
Respect for Authority of Knowledge Claimant	Strong; respect for hierarchy	Weak; demand for evidence

EVOLUTION AS THE CONSEQUENCE OF REPLICATION, VARIATION, AND SELECTION

In the mass media, and even in textbooks, we're often told that evolution does this or does that. Evolution shapes species; evolution causes extinctions; evolution makes plants and animals change through time. These are at best stretching the uses of the word and are at worst misleading. What is misleading is the idea of evolution as a sort of noun, evolution as a machine that builds or destroys. My whole point in this book is to dismantle that false sense of concreteness about "what evolution is." While evolution happens, in some ways there's no *there* there, no such thing as "evolution" with quotes or a capital E.

Rather than a force with any "thingness," evolution is simply the *consequence* of three facts of the natural world. These facts are independent of one another; they don't plan or collaborate, and in fact they can't plan or collaborate, as we'll see. They simply occur, and it is their cumulative result that we call evolution.

Repeating what I mentioned in the introduction, the three facts of the natural world that result in what we call evolution are:

1. The fact that life-forms have offspring; this is the fact of replication
2. The fact that offspring are not identical; this is the fact of variation
3. The fact that some offspring pass more of their genes on to the next generation than other offspring; this is the fact of selection

Evolution is not a thing; it is the *consequence* of these three observable facts. And because evolution is the logical consequence of these three facts, it is no longer debatable; evolution doesn't just happen, it *has to* happen.

Let's see why.

CHAPTER 2

THE FACT OF REPLICATION

So many atoms, clashing together in so many ways as they are swept along through infinite time by their own weight, have come together in every possible way and realized everything that could be formed by their combinations. No wonder, then, if they have actually fallen into those groupings and movements by which the present world through all its changes is kept in being.

—Lucretius: On the Nature of the Universe[1]

We can begin with a question the answer to which seems so simple that you might think I'm kidding. But my whole point in this book is that evolution is in essence simple. So, let's ask: where do living things come from?

Did all these oaks and penguins and other living things just materialize, popping up out of thin air? Of course they did not, you say. Like every farmer or shepherd in history (and prehistory), you know that living things come from parents, whether those parents are acorns, penguin pairs, or what have you. The fact that life-forms come from parent generations is the first fact of the evolutionary process, the fact of *replication*. I would be fascinated to meet with someone who does not accept this first of the three simple processes that together result in evolution.

We can observe replication every day, all around us. Life-forms come from parent life-forms.

We can say more: not only do life-forms come from parent generations, but offspring resemble their parents. Not just on the surface but down to the molecule, life-forms are usually pretty close approximations—replicas—of their parents' basic form. For example, a turtle certainly more resembles its parent turtles than it does some other life-form—say, an apple tree. That is a fact. But why is it true? Why don't elephants give birth to fish? The intuitive answer is that there are different *kinds* of life, and each essentially produces its own kind. Later we'll delve into the natural question of why there are so many kinds of life in the first place, but we need to start by understanding the basics of why life-forms look something like their parents. It's simple enough that life-forms come from parents, but how they do—how replication actually happens—is complex, fascinating, and even near the heart of the definition of life itself. We simply have to understand replication to understand evolution. And we have to begin at the very beginning.

THE ORIGINS OF LIFE, SELF-REPLICATORS, AND THE DNA MOLECULE

The early evolution of life is a fantastically interesting subject. Trying to study it forces us to define life, and being forced to define your foundations is often a fascinating and illuminating exercise. What is life? There are many answers, and their authors are biologists, astrobiologists (if you're going to find life in space, you have to know what you're looking for), and even philosophers. On balance, I think the best definitions are provided by biologist Eli Minkoff and astrobiologists Kepa Ruiz-Mirazo, Juli Pereto, and Alvaro Moreno; according to these and other experts, living things

- **metabolize** (extract energy from and/or convert) substances from their surroundings: this is, essentially, eating, and its necessary result, purging of wastes;
- **move**, converting some of their energy to motion (not necessarily locomoting from one place to another, but moving in some way);
- **respond** to stimulus, such as being touched or detecting a chemical scent;
- **possess a boundary** (for example, cell wall), preventing damage to vitals;
- **grow**, changing size and/or form with metabolizing of nutrients;
- **co-exist** in a community or ecosystem; and
- **reproduce** their basic form in a way that utilizes information and passes information from the parent to offspring.[2]

The final point, reproduction, is the real substance of this chapter. It's very important to biology, of course; it must be, because here it is in the definition of life itself. Living things reproduce; they replicate.

And note that they replicate in a way that passes information from one generation (the parent) to the next (the offspring). This refers to the fact that while some things in nature might seem to reproduce themselves (examples will be given shortly), the way life-forms reproduce is completely different because information directs the replication of life-forms.

If information is so critical to the definition of life, we have to define information. This is straightforward: information is anything that stands for or specifies any one other particular thing or state of things in the universe.[3] For example, white noise, the crashing sound of waves at a beach, contains no discernible information; nothing in it systematically, clearly specifies anything else in the rest of the universe in a way we can recognize. If humans were to pulse electrical signals (for example, Morse code signals) into that white noise, though, and someone listening knows how to read those pulses (Morse code), then we have introduced *information*; specifics now show up in the white noise (like "SOS"). Another way to think of information is that it is always a representation of something else, and not just anything else, but a very *specific* something else. When I draw a "2" on the chalkboard, the particles of chalk are not in the chalk stick anymore, and they're not randomly distributed across the universe; instead, they have been assembled to make the shape our culture has agreed to indicate the quantity "2"—no more, and no less (I did not draw this shape: "3," and the difference is important). I have moved some particles in the universe (chalk) in a way that specifies something else entirely (the quantity of two) from all the other things in the universe those chalk particles might have been arranged to indicate. Information has been generated.

When we think about this, it becomes clear that information is an important property of life and replication. An example can demonstrate how different life is from nonlife. If a boulder rolls down a hill and comes to rest in the middle of a stream, it might be said that the stream "splits" at the boulder, "reproducing itself" to the left and right of the boulder in a very crude way. There are now two somewhat similar streams, one flowing on either side of the boulder, where before there was just the one stream. Compare this "replication" of a natural phenomenon (a stream) to another replication; an oak tree grown from an acorn. In the case of the acorn, a close replica of its parent (a tree) is built by DNA (which we'll come to before long), which carries a lot of very detailed information that directs the episode of replication. These two cases of replication are so different because the stream "self-replicated" only because of the water-impermeable quality of the stone and the physics of hydrodynamics governing the movement of water in the stream. No specific information was involved; the boulder did not specify anything about the universe in the bout of replication, nor did the diverging stream. In other words, no information was involved to structure the relationship between the boulder and the stream. But when a life-form replicates (like the tree from the acorn), the offspring look very similar to the parent because information (not just gravity, as in the case of the water moving downhill past the water-impermeable boulder) specifies that the replication will be (very specifically) a certain way, and not some other way.

Where do we see, in the universe, self-replicating things that have information content? Chemists have for some time known that the distinction between living replicators and nonliving replication can be tough to pin down. The chemical phenomenon of *autocatalysis*—in which a chemical reaction promotes and propagates itself—has been considered, in fact, a precursor to the origin of life. In a recent review of replicators as they relate to evolution, Hungarian theoretical biologists István Zachar and Eörs Szathmáry highlight the large information content of certain replicators as the distinguishing hallmark of life. On a spectrum of information content in replication in nature, from low to high, the authors begin with bread crumbs falling from a piece of bread. These are, in a crude way, replicas of bread, but the crumbs are produced without discrete information (as in the case of my boulder and stream example). While the crumbs may be similar to one another and to the surface of the slice of bread, they bear no resemblance to the whole slice of bread they've fallen from; little or no information directs the formation of the crumbs. A little further along the spectrum, autocatalytic chemical reactions have some information content, but it's pretty simple. Further along the spectrum is the complex coding of information evident (we'll see how in this chapter) in the self-replicating DNA molecule that is the basis of all life on Earth.[4]

All this has led scientists to propose a special class of replicator, the "self-replicator" distinguished by substantial information content. To clearly define successful self-replicators, evolutionist Richard Dawkins has outlined their properties; they survive long enough to self-replicate (*longevity*), they make numerous copies of themselves (*fecundity*), and the copies they make are high quality, capable *themselves* of self-replicating (*fidelity*).[5] Replicators possessing these characteristics can in principle replicate indefinitely, and although they may be "nothing more" than molecules (small, strongly bonded combinations of certain atoms), they have a very different potentiality and fate than that of any non-replicator molecule. High-fidelity self-replicators hum with activity; they are the basis of life. On Earth, humanity so far has observed only a few kinds of self-replicating entity, the most common being the molecule known as DNA.[6]

DNA is a self-replicating entity with high information content. In the correct chemical environment, it will self-replicate. We humans have only started sketching out the details of how this works for about half a century. We've all heard of DNA, and we all know that somehow it's related to reproduction. But to understand just what it is—to appreciate DNA's jaw-dropping, seemingly self-contradictory properties of extreme simplicity and extreme complexity—our best bet is to see just how it was discovered.

DISCOVERING DNA: THE SEARCH FOR THE MOLECULE OF HEREDITY

Over five thousand years ago, somebody in East Asia plucked a wild citrus fruit from a tree and, enjoying its taste, planted the fruit's seeds in the ground.[7] That person was not surprised when, eventually, a tree came up from the seed; it had been known for thousands of years that that would happen. But that person, I am sure, just like all of us at one time or another, must have wondered, "How in the world does a tree come from that seed?"

People have been thinking about this for at least ten thousand years. That's when, worldwide, a number of human populations began to farm, raising food sources (plants and animals) rather than pursuing them.[8] Farmers pay very close attention to very specific properties of plants and animals, selectively breeding them so that today we have thousands of varieties of plants and animals that have been tailored for human use.[9] For example, wild plant foods are normally much smaller than the ones we find at the market, because for thousands of years farmers have been favoring gigantism. Wild oranges are normally less than two inches (five centimeters) in diameter, about half the size of farmed, "giganticized" oranges. That gigantism was easily accomplished by early farmers by simply planting seeds of the unusually large oranges of any crop, over time enlarging the orange. Clearly, those farmers understood the potential of the seed itself. But how, specifically, could that little seed produce an orange tree? Common sense only led so far:

- The seed came from an orange.
- The orange came from an orange tree.
- Plant the seed, and you will get an orange tree bearing oranges and seeds.

We are back where we started. Somehow, *in that seed lay the future orange tree*. Aside from being a gold mine for creative symbolism, that "somehow"—the question of heredity[10]—has been the object of speculation and wonder for many human generations.

We can compress a recurring intellectual pattern in the history of Western civilization now. Early on, the Greeks made some insightful comments about heredity. Over 2,300 years ago, Aristotle (384 BCE–322 BCE) commented on the similarity of children to their parents and speculated that male and female "essences" were combined during reproduction. Some Romans, following Greece (as they often did in matters of the intellect), basically agreed, but their world was conquest, not contemplation, and with the collapse of Rome (around 500 CE) such speculations were largely silenced in the Dark Ages (medieval) period. During the Dark Ages, only fragments of Greek knowledge were preserved. The Library of Alexandria was repeatedly destroyed, rebuilt, restocked, and destroyed. Aristotle was largely forgotten. Superstition overtook knowledge.[11] Women, to the medieval mind, did not contribute to the offspring, they only served as incubators.[12]

And throughout this post-Roman period, of course, the potentiality of the seed was—with the exception of the few heretics—explained as a product of supernatural control; in the Christian faith, God simply made it—like so many other things—so:

"And God said, Let the earth bring forth grass, the herb yielding seed, and the fruit tree yielding fruit after his kind, whose seed is in itself, upon the earth: and it was so" (Genesis 1:11–12).

As ever, such declarations don't leave much to ask about.

It wasn't until the Renaissance (beginning about a thousand years after the fall of Rome) that anyone other than heretics *did* ask.

LITTLE PEOPLE AND GEMMULES

For a long period, sheer desktop theorizing suggested that seeds contained preformed miniatures of the life-form that would grow from the seed. Advances in microscopy (such as the invention of the microscope!) led the Dutch naturalist Nicholas Hartsoeker (1656–1725) to suggest, in 1694, that each human sperm cell—which people had associated with the formation of offspring, in one way or another, for millennia—contained a partly formed, tiny person, ready to incubate in the female womb until birth.[13] Of course, this only raises the question of where *that* little person originated. Today we know—in fact we take it for granted—that the human body develops from a single egg cell that has been fertilized by a single sperm cell. We know that a single fertilized egg cell divides and multiplies, and the cells differentiate into the different tissues of the body, from lungs to hair to eyes, and so on, until the body is built, one multiplication after another.

But, through the early 1800s and into Darwin's time (the 1820s to the 1880s, with his publication of *On the Origin of Species* in November 1859), humanity was just learning all this. Darwin himself wondered how the characteristics of parents were transmitted to the offspring. Grasping a little, he proposed *pangenesis*, a theoretical system of inheritance in which each part of the body (for example, the hair, the eyes, the lungs) produced "gemmules"[14] that migrated from their origins (the hair, the eyes, the lungs) through the blood vessels and collected together in the sperm and egg, and then "built" the offspring body.

To test Darwin's theory, Darwin's cousin, Francis Galton (1822–1911), performed many experiments, transfusing the blood of black rabbits into white rabbits with the argument that if pangenesis were correct, gemmules from the black rabbit blood should show up in the white rabbit offspring. But the white rabbit offspring didn't end up having darker coats, and in 1871 Galton wrote that "the doctrine of pangenesis, pure and simple . . . is incorrect."[15] Now, Galton didn't identify the mechanism of heredity, but by discarding pangenesis he at least ruled out blood-borne gemmules. That's not the sexiest result of an experiment, in the public eye. But in science it's very useful to prove that a particular theory is wrong; it allows you to say, "Well, I don't have to worry about that theory anymore." As we saw in chapter 1, a theory is simply a coherent, logical explanation of a number of observations. If a theory turns out to be false, science moves on to test the next theory. Nobody gets fired (though they may be frustrated); science just moves on.

Experiments such as Galton's were thinning the forest of ideas about the real mechanisms of inheritance. They remained unknown until after Darwin's death in 1882, but while Darwin never clearly understood the true nature of the "somehow" potential of the gemmule (or whatever we want to call the unit of inheritance), it's important to remember that for many reasons, he didn't have to. In fact, because the actual structure and properties of the substance of inheritance were just being explored toward the end of his life, I wonder if the uncertainty associated with new discoveries might have just confused matters. Whatever the case, today we know that Darwin didn't need the details of genetics to understand the essence of evolution, he only needed to know that *replication* was one of the core processes that led to evolution. Darwin knew this. In an examination of why Darwin didn't discover the laws of heredity, biologist J. C. Howard notes: "[Darwin simply] asserted '*No breeder doubts how strong is the tendency to inheritance: that like produces like is his fundamental belief*' and [Darwin] summarizes his position with '*perhaps the correct way of viewing the whole subject would be, to look at the inheritance of whatever character as the rule, and noninheritance as the anomaly.*' Thus, in a sense, the heritability of a character could be treated as a given for the purposes of the general exposition of the theory of evolution by natural selection."[16]

Still, through Darwin's lifetime, improvements in microscopy gave biologists keener eyes and led them, step by step, toward understanding of the unit of inheritance. By the early 1830s, cell division—the actual making of offspring—had been observed and documented, but without an understanding of exactly how it worked. By 1845 it was accepted that all life-forms weren't made just from "living stuff" (hair, muscle, and so on), but in fact were composed of various *types of cells* (liver, lung, hair, and so on). Among these special types of cells were sex cells (the male sperm and the female egg), and not long before Darwin's death, biologists were closely scrutinizing the inner workings of sperm and egg to provide the first clear views of what actually happened in the cells related to reproduction. What they saw was astounding: neither sperm nor egg contained little people or miniatures of any life-forms. What they did contain was even more surprising: they simply contained what looked like minute threads, tiny strands of some unknown substance.

CHROMOSOMES AND DR. MORGAN'S FLY ROOM

In 1882, the multitalented anatomist, teacher, illustrator, and passionate investigator of salamander embryos Walther Flemming (1843–1905) published *Cell Substance, Nucleus and Cell Division*, in which he described strands of this material he had observed in dividing cells. Although he wasn't certain of their function, Flemming felt these particles were important to heredity. He called the "nuclear threads" *Chromatin* (because they showed up when dyed, or *chromed*). Flemming had observed and described what would later be called *chromosomes*, the coils of DNA that ride in our cells and direct the building of everything from muscle tissue to our children.[17]

By 1898, the German biologist Theodor Boveri (1832–1915) concluded that Flemming's chromosomes were indeed not just important to heredity, but critical. Boveri observed that while adult sea urchin cells contained forty-four "nuclear threads" each, their sex cells (male sea urchin sperm and female sea urchin egg) contained only twenty-two nuclear threads each. But when the sperm fertilized the egg, the nuclear threads were combined, such that the offspring had the full complement of forty-four nuclear threads! This proved that the chromosomes were intimately connected with reproduction. Just a few years before Boveri, Swiss biochemist Freidrich Miescher (1844–1895) had even written:

"Inheritance insures a continuity in form from generation to generation that lies even deeper than the chemical molecule. It lies in the structuring atomic groups. In this sense, I am a supporter of the chemical heredity theory."[18]

Right around this time, as Boveri, Flemming, and Miescher were looking directly at genetic material, a number of biologists rediscovered the writings of a previously obscure Czechoslovakian monk, Gregor Mendel (1822–1884). Years earlier, Mendel had reported the seemingly yawn-inducing observation that the characteristics of certain plants—such as the smoothness of a pea's exterior (either smooth or wrinkled)—were not blended at each generation. His thousands of experiments showed that instead, those characteristics were *independently* transmitted from parent to offspring. Cross a tall and a short plant, and the offspring will be not intermediate in height, but either tall or short.

This was critical because in the same way that Galton had knocked off pangenesis, Mendel knocked off another widespread concept, that of *blending inheritance*, in which the parental characteristics were thought (in a very commonsense way) to be somehow "mixed" or "blended" in the offspring. Not so, said Mendel; characteristics are independently passed from parent to offspring, and the appearance of mixing is an illusion that will be dispelled by paying more careful attention to the offspring.[19]

Mendel also showed—and this is what biologists decades after his first report really picked up on—that, somehow, the characteristics of a life-form were not fluid, they were not blurry, there was a substance to them; some sort of grain or concreteness. The implication was clear; maybe Flemming's chromosomes were, somehow, the grains of inheritance.

By 1900, then, blending inheritance was out, and pangenesis was out, and the chromosomes had not only been seen but were known to be the main physical objects that passed from parents to offspring; whatever the "somehow" potential was, it must have been riding on those chromosomes, those fine threads in the cell nucleus. Now things were really moving, and many biologists applied a laser-like focus on Flemming's "nuclear threads," the chromosomes.

From 1904 to 1928 the Kentucky-born biologist Thomas H. Morgan (1866–1945) made a thorough study of the chromosomes of legions of fruit flies bred and sustained in his Columbia University lab, nicknamed the Fly Room. Flies, Morgan reasoned, were cheap, were easy to care for, and bred quickly, so he could examine how their chromosomes were involved with heredity repeatedly (in fact, Morgan was one of the pioneers of the large-scale use of fruit flies to investigate genetics.[20]) Despite complaints from other faculty members about the strong odor of the lab (imagine thousands of flies feeding on mashed bananas), Morgan and his students pursued an intensive research effort, learning vast amounts about the chromosomes. They established that, yes, a species' characteristics did indeed, somehow, ride on those chromosome filaments, and in fact the team began naming specific parts of the chromosomes they found to be involved in the growth of certain aspects of the fly: some chromosome sections were involved in producing "sparkling eyes," other sections, "curled wings," and so on. The first "chromosome map" was published in 1913, showing where, on certain chromosomes, the locations were that controlled certain fruit fly characteristics.

Morgan's work was impressive, earning him a Nobel Prize in 1933. Morgan didn't attend the ceremony (despite his wealth, he was notoriously shoddily dressed and sometimes mistaken for a janitor), but biologist Folke Henschen accepted the prize for him and said:

> The results of the Morgan school are daring, even fantastic, they are of a greatness that puts most other biological discoveries into the shade. Who could dream some ten years ago that science would be able to penetrate the problems of heredity in that way, and find the

> mechanism that lies behind the crossing results of plants and animals; that it would be possible to localize in these chromosomes, which are so small that they must be measured by the millesimal millimetre, hundreds of hereditary factors [sections of the chromosomes], which we must imagine as corresponding to infinitesimal corpuscular elements [bodily characteristics of the fly].[21]

THE END OF VITALISM AND DISCOVERY OF THE DNA CODE

The hunt for the particle of inheritance—the object of endless speculation—was sharpening and quickening; a life-form's characteristics were clearly inherited from the parent as some kind of information encoded in the chromosomes, filaments that ride inside cells. By 1913, T. H. Morgan was freely using the term *gene* to refer to exact sections of chromosomes that related to building certain parts of the body, and by the 1930s, it was well-known that the chromosomes were in fact composed of a long molecule called deoxyribonucleic acid (DNA). This fascinating molecule, clearly (somehow) related to heredity, was largely composed of roughly equal amounts of four chemicals: adenine, cytosine, guanine, and thymine (A, C, G, and T). Somehow, entire bodies—trees, cacti, fish, mice, worms, people—were built by this molecule. Nobody knew how, but at least science knew what to look at.

Based on such chemical discoveries, the *vitalist* conception, which held that life was *vitally* charged with some quasi-supernatural energy—thereby being fundamentally different from all other physical things in the universe—was fast eroding. In 1944, the German physicist Erwin Schrödinger (1887–1961) published an influential book titled *What Is Life?* in which he proposed that life, at base a chemical phenomenon, was distinguished from other chemical phenomena by its information content.

The question was now drawn to a point: how could all the information needed to assemble a fly—or a banana tree or a whale or a human being—ride those microscopic chromosomes composed of just the four basic chemicals? What kind of information was carried on the chromosomes, divvied up into genes? How was that information encoded?

Among Schrödinger's readers were James D. Watson (b. 1928), a physics student at the University of Chicago, and Francis Crick (1916–2004), a British ex-physicist; both eventually went to Cambridge to tackle the DNA problem.

By early 1953, Crick and Watson were working hard—furiously, in fact, in a friendly rivalry with American chemist Linus Pauling—to decode the hard substance of inheritance. "We could not see what the answer was, but we considered it so important that we were determined to think about it long and hard, from any relevant point of view," Crick later wrote, also saying that they worked together because "our interests were astonishingly similar and partly, I suspect, because a certain youthful arrogance, a ruthlessness, an impatience with sloppy thinking came naturally to both of us."[22] The pair also knew that Pauling had just turned fifty and that sooner rather than later the genius would make a bid for a discovery that might earn him a Nobel Prize. There was no time to lose.

Here is what they knew:

- All life-forms are built of proteins.
- The proteins are based on only twenty chemicals, called *amino acids*.
- The DNA molecule is composed of equal parts A, C, G, and T.

How could the four chemicals, A, C, G, and T, encode the production of twenty amino acids that served as the basis for the assembly of proteins into bodies? A large part of the answer must lie in the actual structure of the DNA molecule itself. Like the shape of a key "tells you" the shape of the locking mechanism, the DNA must be shaped such that it could contain a lot of specific information about the amino acids that built the proteins from which life-forms were built. Was DNA a chain, a ladder, a crystal? Based on their extensive knowledge of biology and chemistry, the two worked the problem, one model at a time, sometimes using cardboard cutouts of the DNA chemicals to see how they might fit together.

Many scientific discoveries are accidents; some, though, are genuine eureka moments. One day Watson examined a ghostly black-and-white image of a DNA fragment made by biophysicist Rosalind Franklin (1920–1958). Franklin had been experimenting with inferring the shape of molecules from the way X-rays bounced off of them.[23] "The instant I saw the picture my mouth fell open and my pulse began to race," Watson later wrote.[24] Not long after came the Eureka moment. Playing with his cardboard cutouts of molecules the next morning, and intrigued by the X-shape he'd seen in Franklin's "Photo 51," Watson had a moment of clarity: the shape in the photo, it seemed, was the signature of a twisting ladder shape, the "double helix" so famous today.

When Crick arrived at the lab a bit later, Watson wrote, he didn't get halfway through the door before Watson regaled him with the breakthrough. The next evening the pair worked up a wire model of the molecule, and after some time the structure of DNA—the down-to-the-atom understanding of the particle of inheritance, the thing that made the characteristics of a species, the "somehow" potential itself of an orange seed—stood in Watson and Crick's Cambridge lab. It was a profound moment; DNA, which had built Watson and Crick, was, essentially, looking at itself. It was a moment for a poet. But there was no poet; instead, they went to the nearby Eagle Pub.

The climbing, twisting form didn't look much like anything anyone had ever seen before. "It is a strange model," Watson later wrote in a letter to a friend. Strange or not, the double-helix shape was later verified. The form of the DNA molecule, the thread of inheritance, was known, but in 1954 the pair wrote about the implications with scientific caution: "In conclusion, we may mention that the complementary relationship between the two chains [DNA "rails"] is very likely related to the biological role of DNA. It is generally assumed that DNA is a genetic substance and in some way possesses the capacity for self-duplication."[25]

Still, deciphering the structure of DNA was one of the great discoveries of science; humanity had learned, down to the atom, to understand its own source, the author of its uniqueness and its connection with all other life, because all life—not just humanity—is based on DNA.

In 1962, Crick and Watson were awarded the Nobel Prize, and by 1966, everything was widely confirmed; yes, the double-helix structure was correct, and yes, the DNA molecule was arranged in certain sections, each called a *gene*, that directed the assembly of one of the twenty amino acids (it turns out that genes have other functions, but we can focus on proteins for the moment). Build enough amino acids together, and you get proteins, and build enough of the right kinds of proteins together, and you get bodies: snail, starfish, hawthorn bush, person, sequoia—all constructed courtesy of the twenty amino acids directed by the DNA molecule. The "somehow" potential in an orange seed? It is carried in the seed's particular load of orange-tree-coding DNA, a very long series of As, Cs, Gs, and Ts in the DNA molecule. The same goes for humanity and every other life-form; the only difference is the order of the As, Cs, Gs, and Ts.

THE DNA CODE OF LIFE

What did Watson and Crick's model look like? How did it encode the characteristics of life-forms with just four bases? What follows is a description of how the code works, but if you find yourself boggled by it, don't worry. It's fascinating, but you can understand a lot of evolution without understanding the specifics of the code. So, relax—there's no test at the end of this book. Have a look (figure 2-1 should be helpful), and if it's not your cup of tea, go ahead to the next section, "DNA and You."

The double-helix model proposed that DNA had the form of a long and skinny ladder; so long, in fact, that the chromosome strands that Flemming had observed in 1882 were in fact tightly wound DNA (see figure 2-1). The rails going up the ladder were a kind of "phosphate scaffold," and the rungs going across were built of only the four base chemicals, A, C, G, and T. Watson and Crick further proposed that specific A, C, G, and T sequences along the DNA molecule "coded for" the production of a specific amino acid.

But how? There are only four letters (A, C, G, and T) and twenty amino acids. If the code used only one base (rung) to indicate an amino acid, only four amino acids could be built. If two were used (for example, A, C = some amino acid) there could be sixteen combinations of A, C, G, and T, which is not enough for the twenty amino acids. With three bases (for example, A, T, T) coding for an amino acid, you could have sixty-four combinations, more than you need for the twenty amino acids. Using four bases (A, T, C, C, for example) allows 256 combinations of A, C, G, and T, far more than was needed.

Using ingenious chemical methods, biologists began to insert and delete As, Cs, Gs, and Ts into and from sections of DNA pulled from various species. They discovered that inserting or deleting a single base in a sequence would ruin the whole message "downstream" of the insertion. As J. D. Watson explains in his book *DNA: The Secret of Life*,[26] you can think of it like this: Imagine a sentence with three-letter words: JIM ATE THE FAT CAT. If you delete the first "T," the whole message "downstream" is ruined, resulting in JIM AET HEF ATC AT. Deleting or inserting two base pairs did the same. But insert three base pairs at a time, and the message may survive well enough: JIM ATE ATE THE FAT CAT. From this, it was determined that the sixty-four combinations of A, C, G, and T into three-letter "words" called *codons* was the "DNA code." There was, in fact, redundancy, with more than one way to specify a certain amino acid.

Table 2-1 shows the DNA codons that specify the twenty amino acids; it also shows three A, C, G, T combinations called *stop codons* that act as spacers between the amino-acid-coding codons.

Assemble these acids in different ways, and you have the many thousands of proteins (and other chemicals) built by the DNA that form the body. Only a smattering of the full range of amino acid functions are shown.

Looking at table 2-1, then, we can see that to produce methionine (fifth down on the Amino Acid list) you need the DNA codon ATG; that is the only sequence that will make methionine, and it will not make anything other than methionine. We can (as Matt Ridley suggests in his excellent book *Genome*[27]) envision the genetic code with the analogy of the written English language:

- **Base pairs** = letters (four only = A, C, G, and T)
- **Codons** = words (there are sixty-four possible, for example, A, A, T)
- **Genes** = stories (hundreds of words each)
- **Chromosomes** = collections of stories (a thousand or so per human chromosome)
- **Genome** = all the stories in a set of books (in humans, about twenty-five thousand genes written in twenty-three chromosomes)

Table 2-1. Amino Acids and Their DNA Codons

Amino Acid	Related to/Used in	DNA Codons
Isoleucine	muscle tissue, blood cells	ATT, ATC, ATA
Leucine	blood, muscles, hormones	CTT, CTC, CTA, CTG, TTA, TTG
Valine	muscles	GTT, GTC, GTA, GTG
Phenylalanine	skin pigment, brain chemistry	TTT, TTC
Methionine	hair, skin	ATG
Cysteine	hair, skin	TGT, TGC
Alanine	metabolism (processing foods)	GCT, GCC, GCA, GCG
Glycine	body protein building block	GGT, GGC, GGA, GGG
Proline	connective tissues	CCT, CCC, CCA, CCG
Threonine	nerve, bone, and tooth tissues	ACT, ACC, ACA, ACG
Serine	neural (brain) cells	TCT, TCC, TCA, TCG, AGT, AGC
Tyrosine	hormone production	TAT, TAC
Tryptophan	brain functions	TGG
Glutamine	body protein production	CAA, CAG
Asparagine	central nervous system	AAT, AAC
Histidine	blood, sex cells	CAT, CAC
Glutamic acid	metabolism (processing foods)	GAA, GAG
Aspartic acid	cellular energy	GAT, GAC
Lysine	connective tissues	AAA, AAG
Arginine	immune functions	CGT, CGC, CGA, CGG, AGA, AGG
Stop codons	serve as separators between codons	TAG, TAA, TGA

Figure 2-1 lays out this scheme, and how it builds the human body, visually. Multicellular life-forms, such as a human (A) are composed of a wide variety of cells (B). Each cell contains a nucleus (dark spots in B), which contains chromosomes (C), the "nuclear threads" observed under the microscope as early as the 1800s. Each chromosome (D) is actually a long "thread" of the DNA molecule, tightly wound like a telephone cord

Fig. 2-1. From DNA to Cell Generation

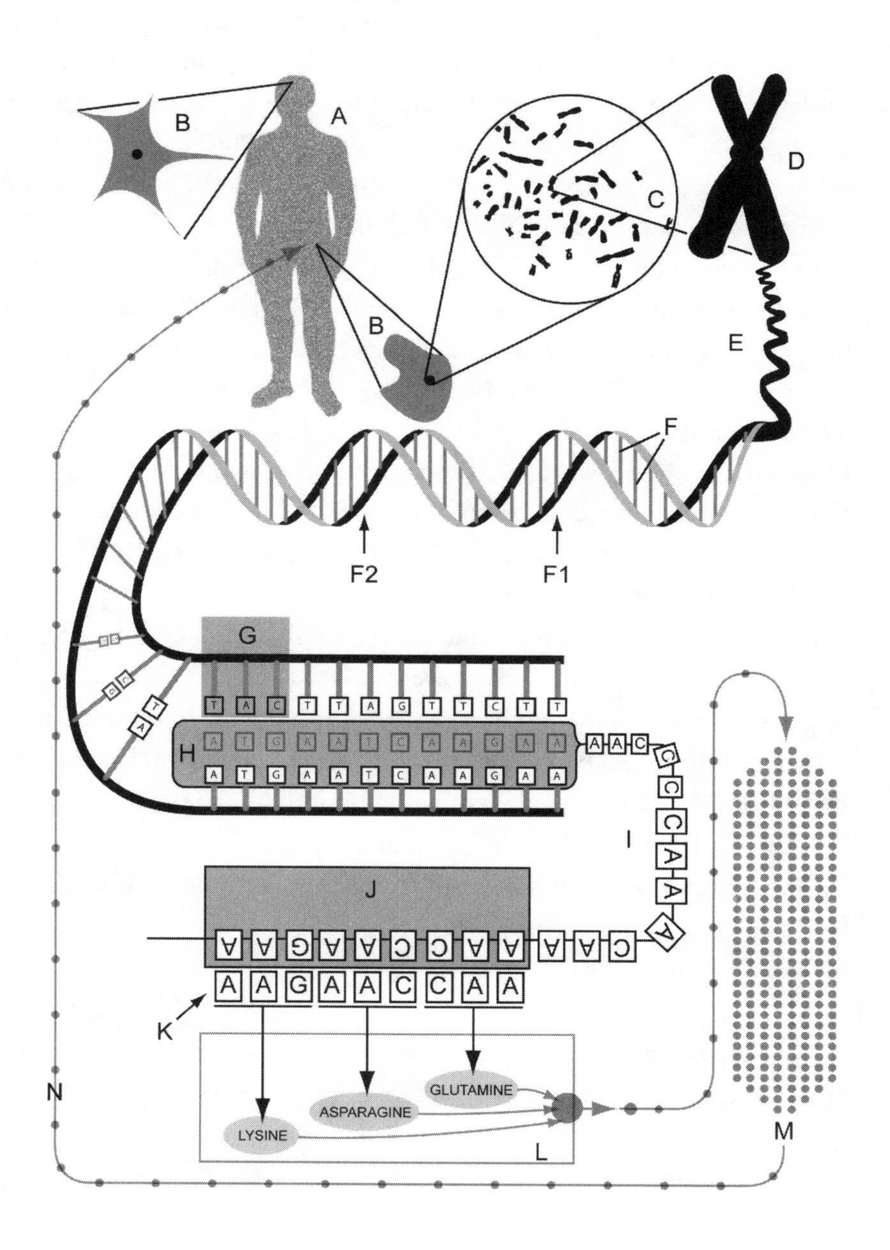

(E). The twisting-ladder shape of the DNA is composed of the phosphate rails (white and black "strips" in E) that connect the rungs (F) of adenine, cytosine, guanine, and thymine (A, C, G, and T), becoming visible just to the left of label (G). Although there isn't room to show a whole gene in this figure, animal genes are normally around 1,200 "base pairs" long, schematically shown as a set of base pairs from (F1) to (F2). Base pairs are arranged in triplets (codons) such as (G), the codon TAC, which specifies the production of tyrosine. When DNA is used to make a protein (for example, to build cells to replace dead cells) an enzyme called helicase (the long gray oval [H]) runs down the length of the DNA (moving to left in this image), "unzipping" the DNA double helix, separating the base pairs. The helicase copies the base pair codons of one side of the DNA ladder (lower and middle rows) and streams them off in a tail-like chain. That chain (I) eventually moves out of the cell nucleus, merging with a ribosome (J), which attaches free-floating chemicals (A, C, G, and T) that will bind only to the appropriate codons, together in the appropriate sequence (K). The resulting amino acids (in this case, lysine, asparagine, and glutamine, assembled in [L]) eventually assemble in the production of proteins, assembled as body cells (M). This figure is schematic, and a universe of detail lies beyond every single thing I've shown. Having said that, what I depict here is enough to understand the system.

Keep in mind that the proteins built by amino acids (such as keratin, used in building hair and skin tissues) don't just form body tissues. Life-forms are made of these tissues, but proteins have many other functions, including the building of

- **enzymes** (these facilitate chemical reactions in the body),
- **hormones** (these carry signals from one kind of cell to another),
- **receptors** (these receive signals from hormones),
- **transporters** (these carry molecules across cell membranes),

- **regulators** (these control rates of certain processes), and
- **switches** (these can activate—or suppress—entire slews of genes).

We'll return to the fascinating world of genes in chapter 8.

DNA AND YOU

How all this plays out, in humanity, for instance, is as follows.

Each human body is built of about one hundred trillion cells differentiated by function; lung cells, liver cells, hair cells, and so on. Each of these *somatic* cells contains the entire complement of DNA required to build a body; if you unwound the DNA in any cell, it would reach about six feet (1.8 meters). Every moment of every day, DNA is making new cells (for example, to replace dead cells) and cranking out millions of proteins. To do this, an enzyme called *helicase* races along the DNA molecule, "unzipping" the DNA of a particular gene. When the DNA is "unzipped," the helicase chemically reacts, copying the A, C, G, T sequence, producing a *transcript*. That transcript eventually detaches from the helicase, moves away from the DNA, and attaches to a *ribosome*, an enzyme that binds the amino acids together, forming proteins.

This dynamic process is characterized by continuous movement and buzzing activity; it is happening at staggering, chemical-reaction speeds, every minute of your life, even now as you are reading.

REPLICATING DNA AND THE ORIGIN OF OFFSPRING

One last thing to consider: we've seen that the body is composed largely of somatic (hair, lung, heart, and so on) cells and that the DNA inside these repairs the body by building proteins and so on. But where did the body itself come from? We know it came from parents, so all we need to do to grasp the essence of replication is see how DNA goes from one generation (parental) to the next (offspring).

The basic answer is that in many life-forms, the body isn't just made of the somatic cells; there are also the very special *sex* cells, also known as *gametes*, in which something very important related to replication goes on.

What happens is this: when the male sex cells (sperm) are being formed, that male's father's DNA—which he carries in his genome, of course—is shuffled with that male's mother's DNA (which he also carries in his genome). This way, his sex cells (the sperm) carry DNA not identical to the male but *a shuffled combination of that male's parents' DNA*.

The same happens in the female: when her sex cell (the egg) is being formed, her mother's DNA and her father's DNA are shuffled (recombined).

We'll see the significance of this shuffling below.

There are two main modes of replication (also known simply as reproduction)—*asexual* and *sexual*. We'll start with sexual reproduction because it's very familiar to us humans.

SEXUAL REPLICATION

Sexual reproduction is familiar because it's the way we and many of the species we see every day reproduce; dogs, cattle, people, birds, and even many plants reproduce sexually. Remember, sexually reproducing (replicating) species' bodies are composed of somatic cells (*soma* refers, in Latin, to the body) and sex cells, called gametes. In males, the gamete is the *sperm*, and in females, the gamete is the *egg*. In sexual replication, when a single sperm cell finds a single egg cell and is absorbed into that egg cell, the egg's cellular wall hardens almost instantly, preventing the admission of any more sperm cells and initiating the processes of *fertilization*. The egg and the sperm cells *each* contain only *half* of the chromosomes required to build the offspring, so neither alone will produce offspring.

Fig. 2-2. An Evening in the Firefly World

But soon the egg and sperm cell's nuclear walls dissolve, allowing the male and female DNA to join. Then the fertilized egg begins to divide—the initial stages of the growth of the embryo. The growth of that embryo is directed by the DNA in its cells, and that DNA of course came from the single sperm cell from the father and the single egg cell from the mother. In humans, it takes nine months for the offspring to be ready to be born, to be ready for its introduction to the environment outside the womb.

The distinction and essence of sexual replication is that it builds a body based not just on the genes of one parent but on the genes of both parents. Not only that, but the offspring is not identical to either parent (nor is it identical to its siblings) because, as we saw above, the offspring's father's DNA had been shuffled before making his sperm cells, and the mother's DNA had also been shuffled before her egg cell was produced. This shuffling, *recombination*, ensured that the offspring carries the *shuffled DNA* of its father and the *shuffled DNA* of its mother.

Recombination is what makes sexual replication so different from asexual replication, as we'll see below. Sex means that the offspring are going to differ from their parents and siblings because they are born of shuffled DNA. This general "difference" is known as *variation*, and we'll see examples of it—and appreciate its significance—in chapter 3.

THE COMPLEX WORLD OF SEXUAL REPLICATION: SOME EXAMPLES FROM THE NATURAL WORLD

The world of sexual replication is extremely complex, so I'd like to take a diversion here, into the wild, where we can learn from meerkats, millipedes, fireflies, and squid.

Why is sexual replication so complex? There are a lot of reasons. For example, many species mate only at very specific seasons, so timing can be critical. In South African meerkats, mating doesn't take place when resources are most abundant (as you might expect). Instead, mating is done in the somewhat leaner times just *before* resources are most abundant, with the result that birth occurs when rainfall is higher and there are plenty of insects and plant foods available; both *gestation* (growing the young in the womb) and *lactation* (suckling the young) are the most "calorie-hungry" times in the life of a female adult meerkat, and she needs all the energy she can get.[28]

Assuming the individual has properly scheduled things, a potential mate has to be found. In a world teeming with life-forms that require certain mate-recognition systems, this opens up the whole world of animal communications that we'll revisit in detail in chapter 7. For the moment, consider that animals communicate visually (seeing), chemically (tasting and smelling), with sound (hearing), and with a wide variety of bodily displays (posturing).[29] Any individual wanting to replicate—and that is a basic urge in life-forms—has to pick through all the communication systems in its environment to find just the right potential mate.

Imagine, for example, the world of a female of one of the nine kinds of the *Photinus* firefly. Figure 2-2 is a schematic representation of one evening in her world.

To attract a mate, our female firefly has to be able to sort through the flashing signals of eight other species before she flashes her own distinct signal to attract one of her own kind. Typical tracks of those other species are shown; some fly higher, some low to the ground. Some flash as they ascend (top line), some flash in repeated sequences (the eight "blips" repeated occasionally on the fourth line down, on the left). A slight variation in the female's ability to perceive these signals, and sort through them, could prevent her from mating successfully.[30]

Things are interestingly similar in the murky world of some central

African river electric fishes; they use distinctive electric pulses, emitted from a special organ in the snout, not only to sense objects but also to identify potential mates of their—and no other—species.[31]

Once a potential mate has been detected, there may be intense competition for access to that potential mate. If the female makes mate choices, this might bring up the whole world of male-male sexual competition, in which males compete for access to the female. Males might actually fight, risking their lives, for the privilege of mating. Female South African squids (*Loligo vulgaris reynaudii*), for example, only accept some male suitors, and this mating behavior takes place in several "mating zones." At medium depth is the agonistic zone, where males encounter one another and fight with their tentacles; when one finally gives up, the victor migrates upward a bit to where he encounters a female in the mating zone. Here, some females are receptive to most victorious males, while in at least one study it's been observed that (for reasons completely unknown to science) females sometimes jet away from a male that has just won a fight. At any rate, once male and female mate, the male may stay around for a while, warning off other males with vigorous fin-flapping and visual displays, or it may move off into shallower water (the immigration/emigration zone), while the female descends to the egg-laying zone. This all seems straightforward enough, but there are mysteries; for example, things in the mating zone aren't so simple; sometimes, small males sneak into this area, closely observe the male and female mating, and, when they separate, dash in to mate with the female. Why this happens, and why the larger males often allow it without opposition, is also completely unknown to science.[32]

Replication isn't only the moment that the male passes sperm to the female, of course; in fact, that's just the beginning. In the squid we've just read about, for instance, the following steps all have to go just right for an egg to be fertilized and then deposited on a suitable egg-bed on the seafloor (where it will, if lucky, develop into a juvenile squid, and then an adult, and so on):

- The egg is extruded from the female's oviduct to the mantle cavity, a cave-like orifice where the male sperm is floating.
- Sperm enter the egg capsule's jelly coat and move toward the egg cell (this can take up to ten minutes).
- The egg is deposited—with perhaps as many as twenty to thirty others—on a suitable site on the seafloor.

Although not much is known about squid mating seasons even today, in some cases this activity takes place over a month or more, with females taking multiple partners and laying up to two hundred eggs.

Even with the mate recognized—and perhaps vigorously competed for—things aren't necessarily over. The act of replication itself can be expensive. One study of mating millipedes (*Alloporus unciatus*) revealed that males burned elevated levels of oxygen during the "precopulatory struggle"—in which they have to uncoil the female from her defensive coiling posture—and females burned similarly elevated levels after copulation, recovering from the struggle. Considering that their mating episodes take around three hours each, and occur repeatedly during the four-month mating season, millipedes had better be in good cardiovascular shape if they're to have millipede young.[33]

Other energetic costs related to sexual replication include the expenses, to the individual suitor, of mating displays (imagine the chest beating and posturing among gorillas), mate defense and territorialism (as when gorillas take on the responsibility, after mating, of protecting their mates), and even parental investment in bringing up the offspring (again, imagine how much time female gorillas spend with their young—it takes years before the young are ready to live independently).

Even when males and females don't come into direct contact—as in many aquatic species that distribute their eggs widely, after which they're fertilized by free-floating sperm emitted by males—nothing is simple. Among crown-of-thorns sea stars (*Acanthaster planci*), oceanic life-forms much like starfish, males must release clouds of sperm at just the right time, and this is best done within about one hundred feet (thirty meters) of the eggs.[34] It might seem that this would never work, with ocean currents moving the sperm hither and yon, but up to 99 percent of the eggs can be fertilized if the conditions are just right. These conditions are found in water shallower than where the crown-of-thorns spends most of its life, so successful migration to a spawning ground is a prerequisite for the survival of the species.

My point in taking these excursions into the squid, millipede, and other worlds is that there's nothing simple about sexual replication. I've described only a few of the processes involved; entire careers could be expended to investigate any one of the facets I've included, and many others I've left out.

ASEXUAL REPRODUCTION

Asexual reproduction is relatively simple and is largely the mode of replication for single-celled life-forms such as the bacteria. In many species, the offspring (referred to as the *daughter* cell, even though there are no sexes in the asexual world) is simply the result of the parent cell dividing in half and releasing one half as the daughter.

There are three main modes of asexual replication: *budding*, *fragmentation*, and *parthenogenesis*.

In budding life-forms the parent cell develops a growth on its body, which eventually *buds off*, effectively releasing the offspring into the environment. For example, the hydra, an aquatic life-form that looks like a stalky plant without leaves, develops a *polyp* that simply drops off the parent before developing into an adult.

In *fragment*-replicating life-forms, the parent breaks into pieces that develop into offspring. If one of a starfish's arms, for example, breaks off, the fragment can, through many cellular divisions, build into an entirely new starfish.

In *parthenogenesis*, while there is a distinction between males and females, the female can in fact self-fertilize the egg (more on fertilization below), resulting in her giving birth without any male involvement. For example, Komodo dragon females (Komodos are very large lizards) occasionally give birth even though there are no males present.

In each of these modes of asexual replication, the thing to remember is this: whether budding, fragmenting, or what have you, what we are seeing in each case is *replication*, the production of an offspring generation not simply by gravity, or as a bread crumb randomly falls from a slice of bread, but informed by specific information. The information is the genetic code that specifies how the amino acids are to be assembled. Assemble them one way, and you get a starfish; assemble them another way and on a different schedule, and you get an oak tree. This is the distinction of life; replicating information, a code.

OTHER MODES OF REPLICATION

There are more exceptions to the rules of the German language, Mark Twain once joked, than uses of those rules. I won't say the same applies to biology, but I will say that we continue to learn about the world of living things, and there is much we still do not understand. Keep your eye on the various scientific journals, and fresh wonders, exceptions to rules, and elaborations on what we already know—and, occasionally, surprises we never expected—come up regularly. For example, *hermaphrodites* are individuals with both male and female sex organs. A lot of snail and slug species, for example, can use either their male or their female genitals to reproduce.

So while the asexual/sexual divide seems cut-and-dried, it doesn't account for every means of replication or every subtlety in the worlds of replication. But the point is that life-forms come from parent generations, by observable processes that we generally understand. And, importantly, the way that replication happens—via DNA-directed processes—is vital, as we'll see throughout this book.

REPLICATION IN EVOLUTION

Sexual replicators, asexual replicators, the wild things that do things differently . . . in the end, they are all still replicators. Life-forms do not pop up out of thin air; they come from parent generations, and the way they do—by DNA direction—is directly responsible for why they look something like their parents.

Examining replication has allowed us to discover and understand the nuts and bolts—the microscopic, particulate essence—of heredity, of how life-forms make copies of themselves. Self-replicators push the information content of life through time.

Self-replicators also build complex bodies, scientifically referred to as *phenotypes* (the root *phen-* refers to something on display or brought to light). We haven't thought much about those phenotypes so far, but in the next chapter we're going to look at them in detail. As you'll see, knowing what you know about the genes is going to be critically important to understanding evolution.

CHAPTER 3

THE FACT OF VARIATION

> **Consider the race of men, the tribes of scaly fish that swim in silence, the lusty herds, the creatures of the wild and the various feathered breeds, those that throng the vivifying watery places, by river-banks and springs and lakes, and those that flock and flutter through pathless woodlands. Take a representative of any of these diverse species and you will still find that it differs in form from others of its kind. Otherwise the young could not recognize the mother, or the mother her young.**
>
> —*Lucretius: On the Nature of the Universe*[1]

We've established, easily enough, the fact of replication; life-forms have offspring, and the development of the offspring from single cell to adult forms is controlled by the DNA molecule. None of this can be denied. Now we come to the second major fact of the natural world that, as we'll see, also cannot be denied. This is the fact of variation; the fact that offspring differ both from their parents and from their siblings. We'll also see that this fact of variation is one of the premises that lead us to the conclusion that evolution is simply the logical and necessary consequence of replication, variation, and selection.

There are many illusions in nature, especially for a species as imaginative as humanity. We humans see a great deal, but we miss a great deal, and what we see is strongly conditioned by our philosophies and our technology. One of the grandest illusions—one that conditioned how Western civilization viewed the natural world for over two thousand years—is that the world of living things is composed of easily defined and internally consistent types of organisms.[2] Now, there certainly are some easily observable types—no one will argue that blue whales are of the same kind or type of life as pine trees—but our civilization's classificatory and typological way of thinking leads us to believe that *within* these types, life-forms are pretty much the same. Without a careful look, one blue whale is about like any other blue whale, and one pine tree is about like any other pine tree. Considering that the early history of biology was largely concerned with identifying and describing all the types or kinds of life—and especially with describing their most distinctive properties in minute detail—it's not surprising that we don't often think much about differences (variation) *within* species, within groups or types of life.

But look closely at any two life-forms of the same generation, or even compare them with their parents, and you will see *variation*, differences between individuals. In fact, when we bother to look closely at life-forms, we break through the mirror-house illusion of uniformity in nature, the illusion of perfect forms. We find that no two life-forms are identical. No two! Look closely enough (and you may have to look at characteristics that are very hard to see, such as an individual's DNA), and you will find that every individual is unique. Even the microbes that produce clone-like offspring show slight variations.

When we think about this for a minute, it makes sense. Have you ever seen two identical pine trees? I mean, really identical, in that there are *zero* differences between the two? Look at figure 3-1; does A or B look more like the real world? Panel (A), in fact, looks strange because

Fig. 3-1. Clones or Not?

we know from experience that there are differences in life-forms; forests don't look like that at all. The real world looks more like (B), a photo of trees in coastal Oregon; no matter how slight, there are differences in every life-form, even among members of the same species.

Even "identical" twins differ in one way or another. In fact, it would be a challenge to find any two life-forms that *are* identical. You can go outside right now and look for identical life-forms, and I guarantee you that if you look closely at any two of a species, you are going to see some differences. Variation is a fact of nature that we can see every day.

Fair enough—life-forms differ. How does this fact fit into the evolutionary process? Why is the fact of variation important? Well, in chapter 2, we established the simple fact that life-forms come from a parent generation; they don't simply pop up out of nowhere; they are born, whether from an acorn or an egg or what have you, but they come from a parent generation. I then indicated that we would see the other two steps, variation and selection, before the evolutionary process falls into place. In the same way, we'll come to exactly why the fact of variation in nature is so critical to evolution—but first I simply want to establish biological variation as a fact.

DNA-LEVEL VARIATION

Whether a variation is very large (for example, a great difference in antler size in deer) or very small (for example, a difference in just a few properties of the genes) is almost irrelevant; any variation has the potential to be important, as we'll see in chapter 4. Let's have a look at one example showing that important variation occurs even at the molecular level.

You may recall Dr. T. H. Morgan's "Fly Room" at Columbia University, where so much pioneering work was done to understand the essentials of heredity. As it turns out, Dr. Morgan's fruit flies (*Drosophila melanogaster*), who subsist happiest on fermenting fruit, have an alcohol problem: the fermenting fruit also produces alcohol in their food, and if they don't have the proper alcohol-mediating enzyme in their digestive system, they become inebriated: "they have difficulty flying and walking, and finally, cannot stay on their feet."[3]

Investigation into this interesting phenomenon showed that the fruit fly has a gene (called *Adh*) that builds a substance called *alcohol dehydrogenase*, which processes the alcohol, preventing drunkenness. If the gene is in proper order—that is, as we saw in chapter 2, if all the As, Cs, Gs, and Ts are in an information-conveying order—all is well, and alcohol dehydrogenase is properly manufactured. But things aren't always well: there is sometimes a significant (but tiny) variation. Remember that a gene is made of codons, which are triplets of base pairs (for example, CCT). As it turns out, codon 192 of the Adh gene has two possible variations: in some fruit flies it codes for threonine (ACG), but in others, it codes for lysine (AAG). The fruit flies with the threonine variation of alcohol dehydrogenase are much more capable of digesting the alcohol and live in better health than those with the lysine variation.[4]

So, empirical studies show that variation in life-forms occurs even at the molecular level. As we'll see in chapter 4, this "tiny" variation can be very significant.

LARGE-SCALE VARIATION

Before we go much further, we need to recall that the phenotype is the organism that's built by the DNA, and that DNA is called a genotype. While there has been plenty of debate in biology over whether the phenotype or the genotype is more important in evolution, the distinction between the two is well established and important. The genotype is the instruction book, and the phenotype is that which is constructed. Evolutionist Richard Dawkins puts it best, saying of the phenotype (the body—be it a cactus or a hummingbird—built by the DNA): "It is the all-important instrument of replicator preservation: it is not that which is preserved."[5]

What is preserved, of course, is the DNA itself, which is sent through time by replication. One way to think of this is that the phenotype is a structure that protects the DNA and moves that DNA around in the world.

The most visible, outward variations in the individuals of a life-form can be called *phenotypic variation*. Variation on the genetic level is called *genotypic variation* (as we just saw in the case of the fruit fly).

The central message of molecular biology is that the genotype builds structures that assemble into the phenotype. That is:

Genotype → Phenotype

The genotype is the instruction book, and the phenotype is what is actually constructed. Information, as we saw in chapter 2, flows only one way. Why the distinction? Because while some phenotypic variation is a result of the DNA code, some other portion of it is determined to some degree by the individual life-form's life history. For example, if an individual has trouble finding food, its body will show that effect in various ways, such as lower body weight. So, that lower body weight wasn't specifically instructed by the genotype, it's a result of the environment the life-form was born into. Two points are immediately relevant.

First, keep in mind that while sometimes environmental factors can modify the DNA (as we'll see later in this chapter and revisit in the section on "mutagenesis" in chapter 8), large-scale, organizational, information-rich changes of the DNA don't occur by environmental control. That is, even if a life-form stumbles into a chemical environment that works a lot of changes on its own genes, it is very unlikely that those changes will result in systematic phenotypic variations, like an offspring fish being born with an insect's wings. The lesson is that while genotype builds phenotype, effects of the phenotype's world do not, as a rule, make direct, organized changes of the genotype; the information flows only one way, as indicated by the arrow in the diagram above.

For example, if I lose an arm in an accident, that does not get encoded into my genes such that my offspring are born missing an arm.[6] While there *are*, as we'll see later in this book, environmental factors that can "edit" the genes during the course of life, and these edits *do* get passed on to the next generation, the nuts and bolts of building phenotypes comes from DNA that isn't altered this way.

Second, the practical point is that biologists observing variation in the wild need to be sure they are documenting not just variation that results from life-course history (like lower body weight from lack of food), but genome-instructed variation, such as when a human is born with six toes rather than the usual five. Since information normally doesn't go from environment to genome in an organized, information-rich way, biologists have to be able to distinguish between genetically instructed variation and environmentally induced variation.

You might be anticipating that this can make the natural world challenging to study, and if so, you're on the right track. Things get wickedly complex when we consider *phenotypic plasticity*, the fact that any given life-form has a range of variation that it might show, *based on environmental factors*. We'll see more about phenotypic plasticity later. For the moment, let's see some more examples of large-scale phenotypic variation.

DISPELLING THE ILLUSION: JELLYFISH—NOT SO ANONYMOUS

As I mentioned, a superficial glance at the world of living things presents the illusion of sameness. How different are those geese, anyway, all those black silhouettes pumping their way across the sky? Well, look closely, and you'll see that they differ plenty. Unless we spend time actually looking at living things, it's easy to get into a kind of "pigeonhole" mentality in which we think of life-forms—species—as perfect types, without realizing how much variation we'd see if we really looked at them individually.

For example, in the top section of figure 3-2 we see a schoolbook diagram of the life cycle of the moon jellyfish (*Aurelia aurita*). At (A) you see the adult form of the jellyfish, which has released a proto-offspring, the *planula* larva (B) after about a week of growth. The planula settles onto a rock (C), where it develops into the *scyphistoma* (D and E) with tentacles allowing it to feed on passing nutrients. Eventually the scyphistoma develop into the *strobila*, a sort of stack of immature jellyfish on a stalk (F) that can bud off offspring (*ephyra*) for several years. When one of the ephyra is released (G), if all goes well, it develops into the immature *medusa* (H), which grows to adulthood (I). Note that this schematic drawing shows a single, largely symmetrical ephyra (G). Seeing such a depiction in a textbook, we internalize a fixed image in our mind: *this is how* Aurelia *ephyra look*. But look at the lower part of the diagram, which shows the actual appearance of twelve *Aurelia* ephyra captured off the coast of California in 1996. Note that no two are identical; there is variation even in this world of "anonymous blobs."[7]

I'm not here to castigate scientific illustrators; naturally we have to make generalizations. What I'm saying is that unless we take a close look at the natural world, we can miss important, subtle differences. And, as we'll see in the next chapter, what we might call "subtle" differences

Fig. 3-2. Jellyfish: Not So Anonymous

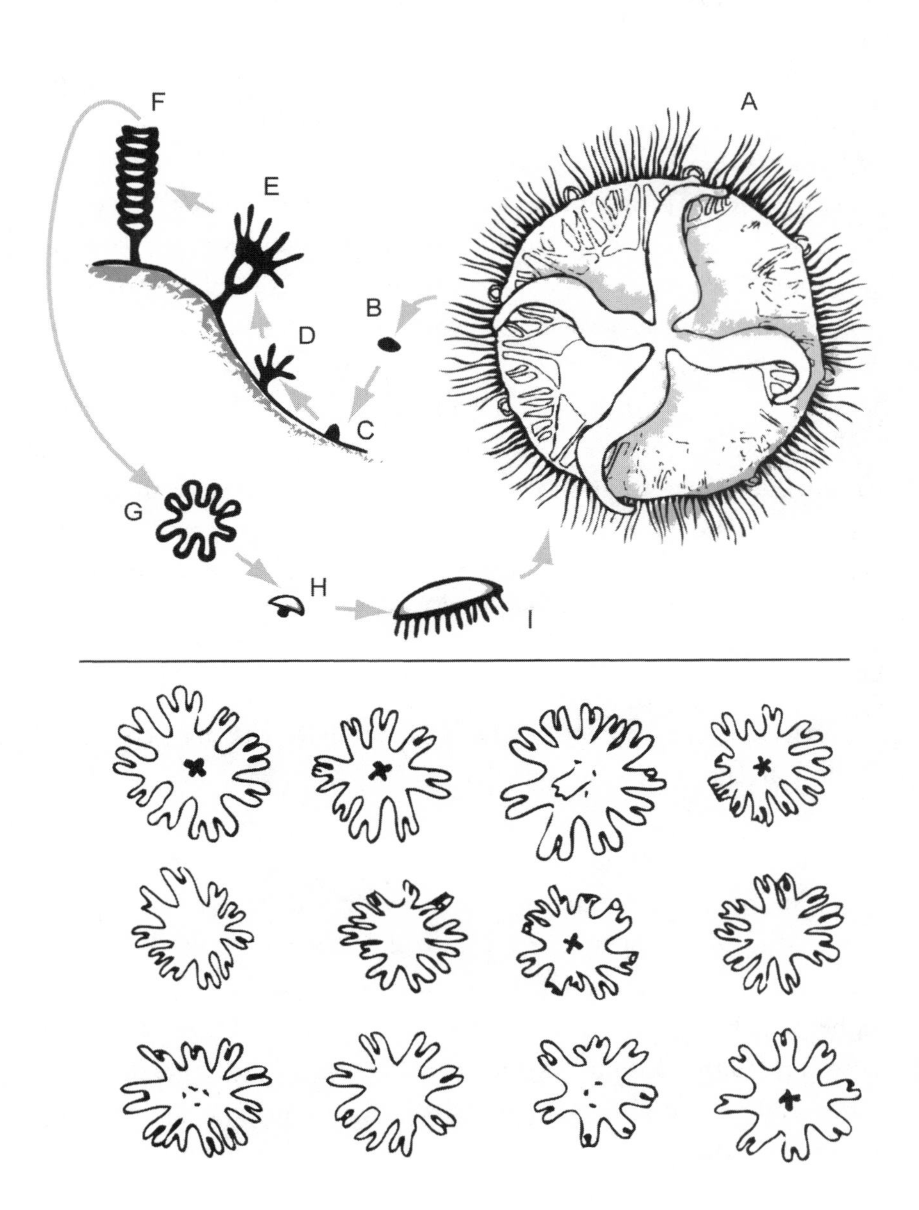

might in fact make all the difference in the life of some other organism. To imagine how, let's turn the tables for a moment. To a box jellyfish (genus *Chironex*), which can examine you with any of its twenty-four eyes, you and your sibling might look more or less the same; both gangly, four-limbed critters clearly out of your environment as you splash around in the ocean. But, of course, you and your sibling are not identical; even if you're twins, you will have differences, and let's say one of those differences is your hearing acuity; you hear very well, but your sibling is nearly deaf. If it comes down to some terrible situation in which you avoid a lethal threat by dint of your superior hearing, then your sibling's poorer hearing acuity—resulting from a microscopic variation, let's say, in the very finest anatomy of the eardrum—would be a genetically significant *variation* in two individuals of your own species (*Homo sapiens*). And that variation would be absolutely invisible to the box jellyfish. The lesson is that when we claim to observe phenotypic variation, we must remember that we might, or might not, be observing all the variations of significance to that life-form.

CONTINUOUS VARIATION: LOCUST BRAINS, BISON CHAPS, AND TURTLE SKULLS—OH, MY!

Variations that can be measured as shades in a gradient, so to speak, can be called *continuously varying*. For example, the height of cinquefoil plants studied in a west-east transect of California showed that they have a range of variation—from a few inches to a foot or more tall—rather than being "punched out" in sizes A, B, and C. They vary roughly according to the altitude at which the seeds take root (figure 3-3 [A]).[8] Another example is seen in the coat coloration of laboratory mice (figure 3-3 [B]).[9] Each of these mice carries the same essential genes for

Fig. 3-3. Variation in Plant Heights and Mouse Coats

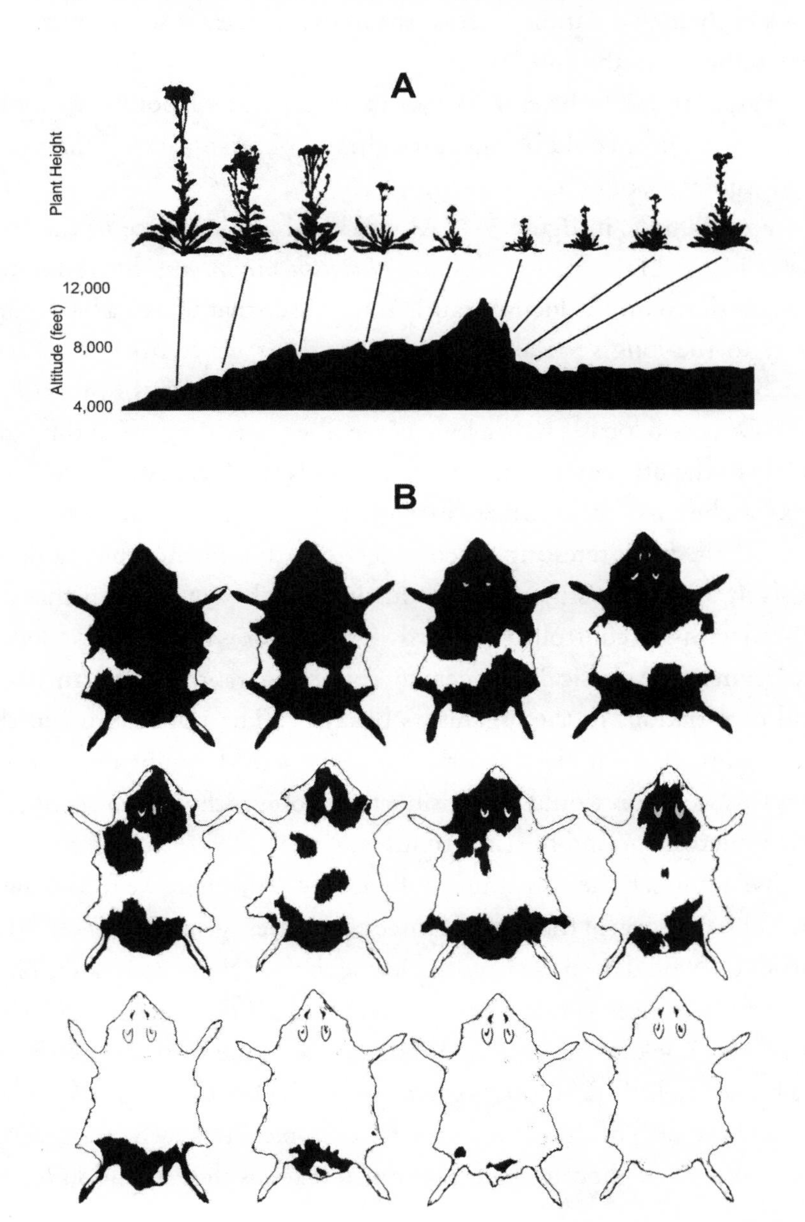

coat coloration, but slight variations in the genes result in slight variations in their coat darkness across the body, so that we see a continuum here rather than absolute "types."

Variation might be easy to spot in a glance at a laboratory mouse's coat color, but it could be much tougher to observe without long-term field studies . . . or even microscopy.

For example, in figure 3-4 (A) you see the appearance of the same nerve in the brains of four locusts (*Locusta migratora*; you're looking straight down on the locust brain). You can see that there's a basic similarity in the long vertical line and a trend for nerve fibers to extend somewhat to the left of the central line, but there's plenty of variation as well. Only two locusts have a long nerve fiber extending off to the right, nearly to the other side of the brain, and only the locust on the far right lacks a cluster of nerve fibers extending somewhat to the right in its upper half. It's interesting to consider that this phenotypic variation exists in the brain tissues, and that instinct—behavior patterns that one is born with—is controlled by "hardwired" patterns of clusters of nerves. Variation in the physical arrangements of neurons (brain cells), then, can lead to variations in the organism's behavior! The implications of this fact will be laid out clearly in the next chapter. My point right now is that this variation would be invisible to anyone without a dissecting kit and a microscope and the time to investigate it.

We see much the same thing in figure 3-4 (B), where we're looking at the right-side view of the skulls of three box turtles (genus *Terrapene*). The turtle's eye would be in the large semicircle to the right of each skull. There are definitely basic similarities, but also some subtle—to you and me—variations. Look at the shape and "pointiness" of the turtles' upper beaks. Each one is a little different. So, here we see the broad similarity—all the turtles have each of the main bones, for example—but each varies a little. Variations we might call "slight" because it takes us close attention to spot

Fig. 3-4. Variations in Locust, Turtle, and Bison

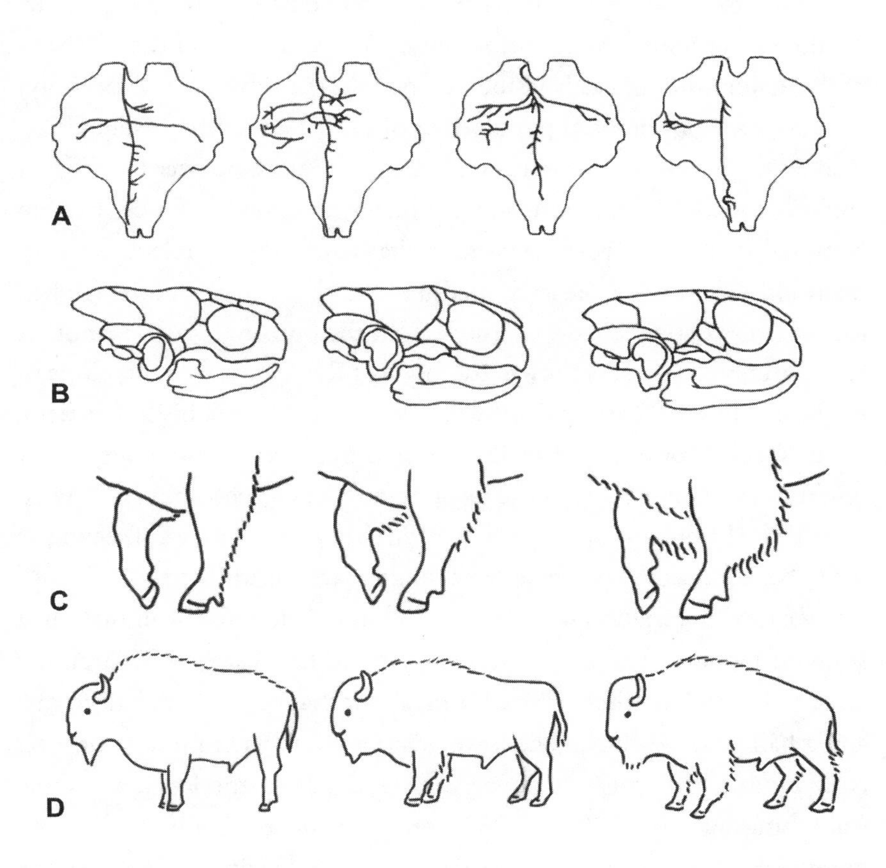

them could make all the difference to the life and health of the turtle, if, for example, it had an effect on how well the turtle could grasp its food (plants, insects, slugs, and snails). Perhaps the more hook-like beak of the center and right turtles is a significant variation contrasting with the "blunter" beak on the left . . . though it could take biologists a lot of work to understand why. The point, again, is that this variation exists, it might be important to the life-form, and it might be difficult to see.

Leaving the laboratory behind and going to the other end of the spectrum, extensive field studies might also be needed to thoughtfully document variations in life-forms. Figure 3-4 (C and D) shows two rows of the major kinds of easily visible (if you take the time to look carefully) variations seen in modern populations of North American bison. Canadian and American researchers identified that these were the most common kinds of variations among bison in Canada's Wood Buffalo National Park. The upper row shows that hair on the forelegs was normally either short (on the left), medium-length (center), or long (right), forming distinctive "chaps" on the legs. In the bottom row (D), another variable characteristic is shown, the angle of the anterior (front) slope of the bison's hump. Once again, if we were to just glance at bison in a field, we probably wouldn't appreciate these differences, even though they might be very important to the bison themselves. One could ask, "Who cares if the bison's hump slopes low or high?" and, again, we'll see in the next chapter that the bison, at least, might care a lot.[10]

We can see variation in all kinds of life-forms, not just in animals. In a study of redbud tree leaves, researchers found that leaves weren't all the same at all; their thickness varied from about that of a human hair to one and a half times as thick. The leaves also varied in their number of *pores* (tiny holes allowing gas transport into and out of the leaf), averaging something between 230 and 330 pores per square millimeter.[11] Who knows what leaf thickness or pore density has to do with the health of the redbud. As we'll see in the next chapter, the point is that just about any variation is liable, at one time or another, to be significant in the survival of a life-form. This is the world of "the difference that makes a difference."

DISCONTINUOUS VARIATION

Discontinuous variation is a little easier to document than continuous variation; it's evident in "either-or" cases. For example, an extreme discontinuous variation might be a fly born without wings—compared to its sibling born with wings—or a person born with five fingers and a thumb rather than with four fingers and a thumb. In figure 3-5 you see in (A) an image from the *Orbis Pictus* (The World in Pictures), a book published as a visual encyclopedia for children in seventeenth-century Europe. On the left you see an instance of discontinuous variation, a person born with two heads, which does occur, if rarely (this image also shows a person with more than the usual number of arms and legs, but it means to show that these variations can exist in different people rather than in just one person). In the middle and on the right you see examples of more discontinuous variation, in the form of very small (dwarf) and very large (giant) people; another common variation in even modern humans is *polydactyly* (*poly*, meaning "multiple," *dactyl*, referring to the fingers), which results in more than the usual number of fingers. Finally, at the bottom (B) you can see the clear difference between highland (left) and lowland (right) common European snails (*Arianta arbustorum*), where the lowland members (on the left) are larger and darker than higher-elevation members (right).

BEHAVIORAL VARIATION

Not only does every life-form vary—even if "only" on the microscopic scale—but what the body *does* can vary as well.

In the flatworm—a biological superstar almost as important as the fruit fly, for many of the same reasons—behavioral variation includes dif-

Fig. 3-5. Variations in Humans and Snails

ferent modes of feeding. Put a colony of flatworms on a "lawn" of *E. coli* bacteria (a favorite flatworm food), and some flatworms will immediately set off on their own; these are the "solitary foragers." Others, though, are "social foragers" that move and feed in groups. One examination of flatworm DNA revealed differences in the *npr-1* gene associated with the nervous system: in the solitary foragers, one of the *npr-1* gene's codons specified the building of valine (GTC), but in the social foragers the codon instructed the building of phenylalanine (TTC).[12]

Now, exactly how this one codon difference resulted in solitary or social foraging behavior isn't well understood. Technically, we can only say that a *correlation* has been shown between the amino acid and the behavior, rather than a *causal explanation*. But other studies have shown—in fruit flies, for example—that similarly "small" variations on the genetic level are correlated with significant behavioral differences, so the inference of some kind of causation is decent. Keep in mind that such studies of codon-level correlations with significant behavioral variations are new, and rare (though increasing in number rapidly),[13] but thousands of studies have shown that animal behavior varies for many reasons, including genetic reasons.

For example, a study of the North Australian keelback snake (*Tropidonophis mairii*) revealed that the environmental conditions in which the snake's eggs nested had important effects on both anatomical and behavioral variation. Specifically, the temperature at which eggs were incubated in nests conditioned, to a degree, the anti-predatory behavior of the hatchlings. Where the temperature of the nest varied a lot from the average temperature of 25.6 degrees Celsius (83.48 Fahrenheit), female hatchlings tended to swim faster than males, despite the fact that males, physiologically, were capable of swimming as fast as females. Anti-predatory behavior—for example, swimming underwater rather than at the surface, and the number of stops a hatchling made during swimming—also varied, depending in some degree on the temperature at which the eggs incubated in the nest.[14] Again, exactly how nest temperature variation affects these behaviors is unknown, but the behavioral patterns are well documented, and explanation . . . well, explanation could come tomorrow. The point here is that some aspects of behavior can be influenced by the environment. These behaviors are in part "hardwired," in that they "ride on the genes" rather than result from learning.

Another example shows just how much "animal personality" can vary, and how that can affect survival—a topic we'll examine in detail in chapter 4. In a study of behavior in the great tit (*Parus major*), a woodland bird common in the Northern Hemisphere, females that were more aggressive and spent more time exploring their environment for food had more offspring during times of food crisis than females that were more retiring. In times when food was more abundant, though, more aggressive females that poked around here and there, exploring the boundaries of their environment for food, tended to get into more conflicts than the more stay-at-home types.[15] Those conflicts led to the more aggressive, exploratory females having fewer offspring than the homebodies.

One fascinating lesson (one that we'll return to in chapter 4) of this example is that certain variations in behavior might be good for an individual at one time (when food is scarce) but not so good at another time (when food is abundant). If this hints to you that something as "simple" as behavioral variation—which is abundantly obvious in all kinds of life-forms—might be very complicated indeed, you're on the right track. A galaxy of complexity lies behind every discovery I'm using as an example. That doesn't mean that we're just skimming the surface—the examples do tell us important things—but it does remind us that, as we explore further, more connections and relationships will be revealed, all leading to complexity, rather than simplicity, in the life and life course of any individual of any life-form. Also, since the effect of behavioral variation—number of offspring, ultimately—was related to differences in food abundance over time, we're reminded to be suspect of short-term studies. Who knows how the trait you're looking at today or this season (such as "exploratory boldness" as in the great tit example) might itself vary, or have different results tomorrow or next season? Field studies should be scaled not to a reasonable time required to attain a doctoral degree but to the realities of a living species,[16] and increasingly this is actually happening, with even multigenerational research projects.

SOURCES OF VARIATION

Clearly, then, everything from the myriad physical properties of a life-form—and from the DNA scale to the whole-body scale—varies. We don't see many clones.[17] Not only that, but behavior also varies. These are facts. Indeed, it would be a challenge to find any two life-forms that are identical in form or behavior. Variation is the usual rule of nature.

Where do these variations come from? In chapter 2 we saw that the self-replicators that are at the origin of life have to be good at making copies of themselves, and I even said that DNA did exactly that; it makes very high-fidelity copies of itself. But very good isn't perfect, and below we'll see how. Also, there are random events that can influence the DNA, such that the offspring differ from their parents and their siblings, and finally there's a powerful biological process related to replication that actively generates variation. Let's see these mechanisms before looking at constraints on variation.

RECOMBINATION

In chapter 2, we saw that there were two basic kinds of replication of life-forms. Asexual replication was a sort of budding-off process, where offspring literally split off from their parents. The main distinction of asexual reproduction, as far as we're concerned in this chapter, is that it normally makes a very close copy of the parent—as far as there are clones in nature (and I'll argue that, actually, there aren't many), they are here in the world of asexual replication. The parental DNA simply, by the mechanism of cell replication and multiplication, makes a very close copy of itself. There is variation in asexually replicating life-forms, but there's less of it than in sexually reproducing species, and the best way to understand why is to look at sexual reproduction's engine of variation, called *recombination*.

We saw that in sexually reproducing species, the body is made of two kinds of cells, the somatic (for example, nerve, hair, muscle) and the sex cells, or gametes. Recombination, we also saw, was the shuffling of the parental DNA during the formation (*gametogenesis*) of those sex cells, the male's sperm cells and the female's eggs. During gametogenesis, the DNA that the male inherited from his mother and father is recombined; the chromosomes (some from the father and some from the mother) come close to one another and exchange bits of DNA. The chromosomes then separate, taking with them any variations that just happened during recombination into the sperm cells. The same happens during the formation of the female's egg cells; her mother's and father's DNA are shuffled, and the egg carries this recombined DNA with all its new variations. Interestingly, male sperm and female egg each have only *half* the DNA required to build the offspring; in this state, they're called *haploid* cells.

What's next? All the palaver of mating takes place—we glimpsed the complexities of that world in chapter 2—and, at long last, one sperm cell makes it to an egg cell. When the sperm enters the egg, the chromosomes fuse once again, *now connecting* the father's recombined DNA with the mother's recombined DNA. The cell that results is called the *zygote*, and rather than being haploid—carrying only half the instructions for building the offspring—it is *diploid* because it carries half the instructions from the father and half from the mother. That full complement of shuffled DNA can now build the offspring. For our purposes, the result of recombination is that the offspring are not identical to their parents, nor are they normally identical to their siblings; they vary. This is in great contrast to what we saw in the world of asexual replication, where offspring bud off from their parental bodies almost like Xerox copies.

So; sexually replicating life-forms, like people, many flowers (remember, their pollen—male sperm, essentially—has to reach the female flower, often depending on insects to carry the sperm), and all our domesticated species, get their variation from recombination. They differ because the DNA that built them has been shuffled and therefore is not a perfect copy of the parental DNA. Shortly, we'll see just how significant variation is to survival and evolution, but first we need to take a quick look back at the asexual replicators.

If asexual replicators don't recombine (shuffle) their DNA, and DNA is so good at making copies of itself, how can there be variation in the world of asexually reproducing species? There are two main ways, and the first is very important to sexually replicating species as well.

MUTATION AND HORIZONTAL GENE TRANSFER

The first is *mutation*. Most of us are familiar with the word in a negative way; say "mutant," and people think of the science-fiction mad scientist who grows a fly's head and arms overnight. But in biology, a mutation is simply a *difference*, whether that difference is *beneficial* (makes life better or easier), *deleterious* (makes life more difficult and the individual less likely to have offspring—like a man growing a fly's head and arms overnight), or even *neutral* (having no real effect on the offspring one way or another). We'll come to the consequences of being born with such variations in the next chapter.

The second reason that asexually replicating life-forms do in fact sometimes vary is *horizontal gene transfer*. This is a fascinating phenomenon by which bacteria, for example, pick up DNA from other species—and not even by replication, but simply by absorbing it into their own DNA. You can imagine how things would go if sexually replicating life-forms began to suddenly incorporate DNA from quite different life-forms and use their newly changed DNA to build their offspring; maybe a person *would* be born with a fly's head (not literally—we'll see why in chapter 8—but I think you get the idea). How bacterial (and other) life-forms pick up these transient fragments of DNA, we'll also see in chapter 8. For the moment, it's enough to know that there are known processes by which even asexually replicating life-forms, which I've (perhaps unkindly) referred to as Xerox copies, do actually vary from their parents and siblings.

Whether it happens sexually or asexually, then, variation is a fact of the natural world. We've seen that there are all kinds of variations, and also that they have all kinds of sources, and we've had a glimpse at their consequences, which are the subject of the next chapter. Table 3-1 summarizes some of the main sources of genetic variation.

CONSTRAINTS ON VARIATION

While we've seen that a lot varies between offspring and parent, and between offspring and their siblings, it's pretty obvious that the range of variation is not infinite. Elephants may have different-sized ears, but they don't have human-sized ears, normally, and they certainly aren't born with wings. In fact there are some pretty strict limits to how much an individual will vary from its parents or siblings. In a very real sense—that we can see every day, all around us—the apple doesn't often fall too far from the tree.

Table 3-1. Sources of Variation

Source	Mechanism	Example
Mechanical Properties of DNA	Temporary mechanical deformation of DNA molecule affects the base pairs.	"Wobbling," "buckled," or "slipping" DNA strand may result in a T being replaced by G.
Mutagen	Energy or chemicals directly encounter and alter DNA.	X-rays can "link" two T (thymine) base pairs, where normally there is only one. Radiation energy can break an entire DNA strand in two. Alkaline conditions can *separate* DNA "rail" strands.
Insertion or Deletion	Insertion or deletion of base pairs that are not whole codons (not divisible by 3) into the DNA sequence.	Deletion of a single "A" (the first "A" below, for instance) so that rather than JIM ATE THE FAT CAT, the triplet-based codons read gibberish: JIM TET HEF ATC AT.
Recombination	"Shuffling" of genes before production of next generation.	Male DNA is shuffled before formation of sperm cells; female DNA is shuffled before formation of egg cells; offspring inherits shuffled characteristics.
In-Migration	New genetic material is introduced into population.	A population begins mating with newly discovered population of same species, "importing" new genetic material into the "gene pool."
Horizontal Gene Transfer	New DNA is taken into the organism not by replication but simply by absorbing it from the environment.	Unrelated or distantly related species transfer DNA back and forth; largely known in microbial species (today).

EVOLUTIONARY HISTORY

One of the main restrictions on variation is a life-form's evolutionary history. As we saw in chapter 2, a life-form is built of the DNA it inherits from its parent (in asexually replicating species) or parents (in sexually replicating species). Since that DNA came from the parent generation, which built their basic body form, and since they inherited their DNA from a generation that built a similar body form—and so on, back for who knows how many generations—it's clear that there is a basic and largely unique DNA code for the basic structure of each basic form of life. The three billion or so As, Cs, Gs, and Ts of the human genome (for example) cannot be shuffled willy-nilly if the offspring is to survive; the sequence matters. Large-scale and rapid reorganization of the information content of a species' DNA code just can't happen in any way that preserves enough organization to build a healthy offspring. For this reason, many variations are relatively small, on the order of inserting or deleting a base pair here and there, producing, normally, relatively small variations on the basic theme.

Planting crops and breeding animals, as much of humanity has done for thousands of years to survive, is a good proof of the reality of the evolutionary history of a life-form in generating the next generation; if breeding two sheep turned out, randomly, hyenas or bats (let's say), we probably would have given it up. So: a cactus has DNA that codes for basic cactus-ness, and a cat has DNA that codes for basic cat-ness. That's clear.

Having said that, before moving on I must communicate one new and significant fact. Despite what I've just said—that there are basic *genomes* (sets of genes specific to one kind of life, for example, the dog genome versus the sea slug genome) for different kinds of life-forms—it has been found that there are a few hundred "immortal genes," shared among all animals, that are responsible for the essential shape of a given animal. It used to be thought that, generally, different kinds of animal bodies would need very different genes, which intuitively makes sense. But, as evolutionary biologist Sean B. Carroll has pointed out, all animals, "no matter how different in appearance, share several families of genes that regulate major aspects of body pattern."[18] We'll examine the jaw-dropping (I really mean that!) significance of this fact in chapter 6, but for the moment, I want to emphasize that while it is the case, it doesn't dilute the significance of the history of a given life-form.

The significance of a life-form's evolutionary history is well conveyed in the term *heritage*. Heritage refers to a sort of organic whole of *what is inherited*, and what is inherited, as we've seen, is the specific (but shuffled) DNA of the parental generation. Since the DNA is, largely, specific to a given life-form, the life-form born of DNA will more or less resemble the parent.

One of the more stimulating metaphors for this DNA heritage is that of the *bauplan* (*bau*, "to build"), a German word referring, essentially, to a blueprint.[19] This concept was introduced to modern biology by evolutionists Stephen Jay Gould (1941–2002) and Richard Lewontin (b. 1929) in 1979.[20] In their very influential critique of biology at large, Gould and Lewontin chucked, as it were, a hand grenade into what they saw as a fossilizing world of biological thought. Biologists, Gould and Lewontin argued, were too self-confidently explaining every single characteristic of a given life-form as being perfectly built (adapted) for a specific function; the forearms of *Tyrannosaurus rex*, they argued, weren't necessarily "for" anything—as many biologists suggested—instead, they might just be the *vestiges* (remainders, or evolutionary holdouts) of some ancient purpose. They existed, however, in *Tyrannosaurus rex* because of an evolutionary heritage; the *Tyrannosaurus* bauplan—its essential DNA code—directed, essentially, the building of a basic *Tyrannosaurus* form, even though the front legs lost their original function over time as *Tyrannosaurus* itself changed.

You don't have to go to the fossil beds, though, to find evidence of an ancient body plan. Although human ears aren't very mobile (and we certainly don't live or die depending on how well we can move our ears), there are muscles associated with our ears; these are *vestigial* characteristics, characteristics seen in the present that are carried on through time because of evolutionary heritage (bauplan), not because of their function today. And things can get even more complex when we consider the phenomenon of *exaption*, in which a characteristic that used to serve a certain function lost that function but was preserved through time and, later, becomes useful for some new function! Why would a characteristic persist through time if it's no longer used? There are a few reasons, but one important one is that it might be genetically linked to some other characteristic that's still being used, so it can't simply vanish.

The bauplan concept reinforces the basic fact we examined in chapter 2, the fact that offspring look generally similar to their parents and siblings. As we've seen in this chapter, however, they are not normally *identical* to their parents and offspring. Again, we'll come to the significance of these variations in the next chapter. For the moment, let's see some more limits to variation.

PHYSICAL CONSTRAINTS

There are realities of the world that any life-form must face. If a bird is to fly, its wings must provide so much lift, and the bird itself mustn't weigh more than so much, and the bird must have so much muscle tissue to work its wings and other structures. Physical constraints like these—and there are certain ones for every species of life—form a kind of "hard line," a boundary beyond which variation can't go if the individual is to have any success at survival.

A good example is hummingbird hovering. This kind of flight is the most energy-demanding in all of the flying life-forms, because the bird is held in the air entirely by its own effort (flapping its wings) rather than with any help from air currents or lift generated by the bird flying forward. To hover, hummingbirds have to have very powerful wing muscles (the pectoral muscles, which provide much of their strength, account for 25 percent of a hummingbird's weight) but they also have to have low body weight. One study found that smaller hummingbirds had smaller wing muscles; the muscles were enough to allow hovering but, because of the weight restrictions of hovering, were only large enough to provide enough energy for hovering, and not even a tiny bit more. Another study found that hummingbird wing length and weight were strongly correlated; that makes sense because a wing simply has to have certain properties if it's going to work; any variation could be disastrous.[21]

Of course, sometimes individuals *are* born with characteristics that exceed this hard line (or they grow out of the boundaries during their lifetime), but these individuals aren't likely to pass their genes—genes for building a body that exceeds the hard line—on to the next generation. We'll come to the issue of why in chapter 4, but for the moment, the lesson is that physical realities of the world often constrain variation in life-forms.

DEVELOPMENTAL CONSTRAINTS

We've been looking at variation in the fully developed life-form, but we know that life-forms grow from an earlier stage; from just a few cells, for example. It turns out that the facts of the development of a life-form can also put severe limits on the kinds of variations that can occur—a better way to say this is that variation that might occur during critically important developmental stages won't be tolerated, the life-form carrying such variation won't survive, because of the interruption of crucial developmental periods.

For example, one study of early developmental sequences showed that in both the common mouse (*Mus musculus*) (which has about twenty-six stages of development) and the zebrafish (*Danio rerio*) (which has about fourteen stages of development), far less variation was "tolerated" in the early stages of development than later.[22] Those early stages of life development (as might be expected) simply included too many critical processes (such as development of the basic organs) for individuals carrying such variation to survive for very long. Biologist Louis Wolpert has stated that "[i]t is not birth, marriage, or death, but *gastrulation* [an early animal developmental stage], which is truly the most important time in your life."[23]

Even after early development and birth, the phenotype continues to grow and change. For example, in Lubber grasshoppers (*Romalea microptera*), once the adult female has biologically "committed" to laying eggs, a number of hormonal changes occur that strongly structure her hormonal development and her egg-laying behavior, and variations in this period are rarer than before it (when she was juvenile). In this very complex case, the point is that how a life-form changes over its life course both (a) varies and (b) might constrain the kinds of variation that can occur.[24]

DNA REPAIR

In addition to these mechanisms is the fact that the molecules that build life-forms are, when damaged, not necessarily passive. As we saw in chapter 2, DNA doesn't just direct the assembly of proteins that build into body cells. It also builds many enzymes, chemicals that facilitate other chemical activity. Some of these enzymes have been observed to repair damage to DNA. For example, when a series of base pairs—the As, Cs, Gs, and Ts we looked at in chapter 2—are damaged, special enzymes actually snip the DNA rails on either side of the damaged base pairs and then excise the broken piece. Then the enzyme uses the bases on the *other* side of the ladder (see figure 2-1 to recall that any rung is made of pairs of A, C, G, and T that match *only with their counterparts* on the other rail) to manufacture the *correct* base pairs, which are then inserted in the correct position on the flawed side of the DNA.

Other repair processes can reconnect the rails of the double helix when one or both are broken, as can happen when the rails are exposed to certain kinds of radiation. To date, over 130 genes involved in human DNA repair have been identified, and we can be sure more will be discovered as the human and other genomes are explored.[25]

The significance here is that DNA repair can indeed reduce the amount of potential genetic variation in a given life-form.[26]

POPULATION-LEVEL CONSTRAINTS ON VARIATION

The limits to variation we've looked at so far are all limits on how much variation can occur in a given individual of a given life-form. As we'll see in the next chapter, we also have to consider, for many reasons, the entire population of a life-form. Such a group is often called a *species*, which we'll investigate in detail in chapter 5. For the moment, keep in mind that there are several general population-level processes that can also reduce the genetic variation seen in a population of a life-form.

DRIFT, THE FOUNDER EFFECT, AND POPULATION BOTTLENECKS

The first of these processes is *genetic drift*, and, once again, I'm afraid the term is somewhat misleading. Genetic drift has been defined many times, but a good consensus definition is that drift is *randomly based change in the gene pool*; that is, it's not the result of selection (we'll return to selection in the next chapter), or other processes that systematically change the species, it's the result of some random occurrence. A good example of pretty obvious drift is the *founder effect*, in which a limited population of a given species provides the original genes that later reproduce and flourish. For example, if a bird species occupies all of some continent, but some small population of that species breaks away (let's say a storm carries them to a distant island) and founds a new colony, the founders' genes will condition the characteristics of the subsequent population's gene pool. If there's no significant breeding back with the original, continental population, the founder effect could be very important.

Many studies of the founder effect are, in fact, carried out on islands where continental species have in fact arrived by long-distance migration or by sheer accident. One review of the genetic diversity (variation in the gene pool) of over a hundred island mammal, bird, reptile, insect, and plant species showed that island species had around 30 percent less genetic variation than their continental counterparts.[27]

In a similar case, in 1912, Ohio State University student Freda Detmers planted a single pitcher plant (*Sarracenia purpurea*) on Cranberry Island, a seventeen-acre floating bog in Buckeye Lake, Ohio. Pitcher plants had not previously inhabited the island. By the mid-1970s, the pitcher plant population was flourishing (over one hundred thousand plants), and botanists K. E. Schwaegerle and B. A. Schaal had an idea to test the founder effect concept: compare the genetic diversity of the

Cranberry Island pitcher plant population to that of other pitcher plants in the area (five nearby states). The simple study revealed that indeed, genetic diversity in the Cranberry Island pitcher plant population was significantly lower than in the surrounding populations.

The founder effect can also be played out anytime a population is severely reduced but then survives that reduction. For example, when plant ecologies break up (for whatever reason), and plant populations are reduced, the resulting *bottleneck* can drastically reduce genetic variation.[28] Today's cheetah populations have a very low genetic diversity; they're all very similar, and it's suspected that this resulted from two severe population crashes in their history, about ten thousand years ago and sometime in the last century.[29]

MIGRATION

New genes can come into a gene pool with immigration, but (since we're talking about constraints on variation), think about out-migration, the movement of genes out of a population of some species. For example, in gorilla society, males are "kicked out" by the silverback (the dominant male) when they get to be sexually active and are therefore a threat. In these cases, whatever variation those males carry in their genes will not become part of the local population.

SELECTION

Another factor that can reduce the amount of variation in a population is natural selection, a phenomenon we'll examine in detail in the next chapter. For the moment, it's enough to know that when an individual

Table 3-2. Constraints on Variation

Source	Mechanism	Example
DNA Repair	DNA-level processes restore original order of DNA code when it is altered.	Enzymes snip out sections of damaged DNA and replace them with correct sequence.
Evolutionary History/ Bauplan	DNA sequence "settles" into complex, integrated code with a kind of "inertia."	Meaningful, integrated, large-scale sequences cannot simply "pop up" in the genome. Offspring will be built largely on the same body form as their parents and siblings.
Developmental Constraint	Critical periods of growth tolerate little variation.	Variation during growth of circulatory pump (heart) or neurons (brain) can be disastrous for the organism.
Founder Effect	A small "founding" population has low genetic diversity, and later generations will also have low genetic diversity.	Island species originating on mainlands typically have low genetic diversity.
Small Breeding Population	Small population has low genetic diversity, resulting in abnormal offspring.	Rabbit ranchers know that high genetic diversity is needed to avoid lethal mutations in rabbit populations. All human cultures have taboos regarding incest.
Selection against Variation	Variations born into environment are not passed on to the next generation.	Wingless fly; blind snake.

of a given species fails to pass on its genes—its genetic bauplan—to the next generation by having its own offspring, natural selection has effectively deleted whatever variations that individual might have contributed to the gene pool. When a deleterious variation (in behavior, anatomy, development, or anything else we've seen in this chapter) comes up, it's unlikely to become common in a population, and that is a reduction in variation in the population. This is called *selection* (the focus of the next chapter), and it, too, can constrain variation.

VARIATION IN EVOLUTION

Well, you get the idea. We can observe variation directly, and whether it's discontinuous or continuous, whether it's on the DNA level, or that of the physical characteristics of the individual organism, or even in the organism's behavior (and perhaps all three), life-forms vary. This cannot be challenged by a rational person. Not only is variation all around us, we know some of its sources and some of the constraints to variation. We know why it happens and what prevents it from happening.

So far we've seen that replication is a fact and that variation is a fact. These variations, we'll see in the next chapter, can be the "differences that make a difference" in the world of natural selection.

CHAPTER 4

THE FACT OF SELECTION

> **And many races of living things must have died out and been unable to beget and continue their breed. For, in the case of all things, either craft or courage or speed has from the beginning of its existence prohibited and preserved each particular race.**
>
> —*Lucretius: On the Nature of the Universe*[1]

We've seen now that life-forms are high-quality replicas of parent generations because of the information-rich content of DNA. But we've also seen that they're not perfect copies, even though they may look it on the surface; there is variation. These are both observable facts. Now we come to the third observable fact of the natural world involved in the evolutionary process: the fact of selection, a fact that is a simple consequence of the fact of variation.

NATURAL SELECTION

It's easy to understand why the phrase *natural selection* is often misunderstood. First, the word *selection* implies *a selector*, someone or something making a decision, *a selection*. This chapter will show—starkly—just how false an image that is. Second, the term *natural* implies that there is unnatural selection. Humans definitely direct the evolution of some plants and animals (by farming), but humans are products of the natural world, so the word *natural* is unnecessary.[2]

We're left with *selection*.[3]

Selection refers to the fact that *not all members of a population will have the same number of offspring.* Some will have no offspring, and others will have few, and others will have many. This is observable around us, every day. It may seem hard to believe when we consider humanity; most of our offspring (at least among those lucky enough to live in developed countries with good health care) seem to survive well enough. But look a little beyond our own world, and you see the fact of selection everywhere.

Many forms of life don't produce just one, or two, or three offspring but many thousands or even millions; and most of those offspring do not survive. For example, it's estimated that only one in a thousand loggerhead sea turtle offspring survives long enough to reproduce.[4] Look at acorns: one study found that 99 percent of the 23,000 acorns in a California forest survey area were rotten or otherwise unable to produce an oak tree.[5] And the list goes on: only half of all polar bear cubs survive to their average reproductive age (beginning around five years)[6]—about the same survival rate as ravens in one Mojave Desert study.[7]

In the terminology of evolution, what we're looking at here is selection: the fact that more offspring are born than survive long enough to have their own offspring. And even of the ones that have offspring, some have more than others. Why?

The answer lies in what we have learned so far. In fact, it's a *consequence* of what we learned in the last two chapters. Recall that (as we saw in chapter 2) offspring are essentially copies of their parents, built by DNA. But in chapter 3, we saw that, for a number of reasons ultimately rooted in DNA, the offspring are not identical copies of their parents (or their siblings). Rather, offspring vary—sometimes a little, sometimes a lot—and variation, not cloning, is the rule of nature. Now we come to selection. Why do only a portion of offspring survive to have their own offspring? Which *do* survive? We can answer both questions with one observation: the offspring that survive are the ones that are better suited to their environment than their siblings. These are the individuals with variations that give them a slight edge in survivability and a slight increase in the likelihood that they will have offspring. This probably reminds you of the famous phrase "survival of the fittest," and this leads us to the term *fitness* in general. A quick look at the fitness concept will make natural selection instantly clear.[8]

FITNESS

The term *fitness* has many appropriate uses; one recent review of the concept lists twenty-eight slightly different meanings, and entire books have been written on the subject.[9] But, generally speaking, *fitness* refers to an individual organism's likelihood of passing its genetic material (the DNA that made *it*) on to the next generation.

Passing on its genes means surviving long enough to find a mate and then successfully mating such that they produce a healthy offspring. Individuals that have a greater chance of doing this than their siblings (or any others of their species) have a higher fitness "score."

Evolutionary biologist John A. Endler provides five short definitions of the fitness concept; note that each one ultimately has to do with the probability of having offspring and/or the number of offspring produced:

1. A measure of mating ability
2. Relative contribution to the gene pool

3. General "adaptedness" of a phenotype (body built by DNA) to its environment
4. Capacity for adaptation to environmental variation
5. Likelihood of leaving genes that will persist long into the future[10]

I think Richard Dawkins put it most clearly: "The fitness of an organism is its success as an ancestor or, according to taste, its capacity for success as an ancestor."[11]

Fitness, then, is something of a probability figure. I like to think of it as a "cosmic wager" on your genetic prospects for the future, the likelihood that you will have offspring.

What makes for fitness? There is no single answer. For a dolphin that uses echolocation to find food, having slightly better hearing than a dolphin swimming right next to you—because your parents passed on genes that made better hearing structures in your head—would mean a slight bump up in your fitness score because you will be better able to find food, leading to better health, which leads to a greater likelihood of finding a mate, having offspring, and thereby ensuring that the DNA for such fine hearing persists in the population.[12]

We'll see plenty of examples of variations in fitness throughout this chapter (and this book); the point right now is that fitness can be thought of as a probability figure indicating your likelihood of having offspring, and that it is based on variations in a population—the kinds of which, and sources of which, we examined in the last chapter.

To bring us up to date; replication makes life-forms, but those life-forms vary, and their variations result in different degrees of fitness. This is all crystal-clear and observable in nature all around us, every day.

Two issues need to be dealt with before we move on.

First, you may have heard or read critiques of evolutionary theory claiming that that the term "survival of the fittest" is a *tautology*, a statement that uses circular reasoning. An example of a tautology is "All bachelors are unmarried," and it is circular reasoning because we define bachelors to be unmarried. The same problem, it is claimed, applies to the phrase "survival of the fittest." Why is this a problem? Because we define the fittest as those who survive. This critique, however, is quickly dismantled by evolutionary biologist Eli C. Minkoff: "According to this objection, the 'fittest' are recognized only by the fact that they have survived, and so 'survival of the fittest' becomes 'survival of the survivors,' which is a tautology. But Darwin never spoke in these terms, arguing only that some variations would confer 'advantage' and that others would be 'injurious.'"[13] In other words, observing that some individuals have greater fitness than others, giving them a greater chance of having offspring, is not part of a tautology; it is simply an observation. The great biologist Ernst Mayr (1904–2005) also made quick work of the tautology argument, saying that Darwin, rather than using the term "survival of the fittest," should have simply said "survival of the fitter."[14] Who are the survivors? The ones that were fitter than their companions, as is easily observable in nature every day (for example, a fly born without wings is less fit than its siblings born with wings). That takes care of that.

Another issue related to fitness is a common misconception: that when we invoke the phrase "survival of the fittest," we are characterizing the natural world as dominated by the violence of predation and fighting. This is also easy to dispense with. As my co-author Charles Sullivan and I wrote in *The Top Ten Myths about Evolution*:

> "Survival of the fittest" is the most commonly used phrase drafted into everyday speech from the theory of evolution. Flipping through television channels, we see a lion bearing down on a gazelle, a boxer pummeling his opponent, bighorn sheep clashing horns: we nod and

> smugly think, "There, see? *Survival of the fittest*; the order of nature." And it seems clear enough: for all we can tell, the strong *do* survive. It would be crazy to think otherwise, considering what we've learned about the natural world from mass media.
>
> But mass media, of course, is about drama and unfolding stories, and every dramatist knows that without conflict, you have no story. And so the natural world has been dressed up as a vast and violent landscape of competition, the ultimate reality show, one where real blood can be shed. Can the antelope corner tighter than the lion? Can the species survive? Will the "balance of nature" be upset? Television has taken Tennyson to heart, portraying nature as "red in tooth and claw," a world of savage predation, where survival of the fittest is the primary law.[15]

I don't want to spend much time dispelling the common misconceptions of the evolutionary process in this book; if you're interested in them, I invite you to read *The Top Ten Myths about Evolution*. What I want to do in this chapter is focus on selection: exactly what it is, how it plays out in the natural world every day (observably, all around us), and how and why it is one of the core natural processes that results in evolution.

We've seen that life-forms vary and that those variations result in differences in fitness. Clear enough. Returning to the central issue of this chapter, we come to selection.

Selection, as I mentioned, refers to the fact that not all the offspring of any given generation survive to have their own offspring.

We asked why this was; now, armed with a better understanding of selection and fitness, we can answer this question: *why* don't all the offspring survive? Because their variations give them differences in fitness. Examples throughout this chapter will make this clear. Before looking at some examples, we need to understand two more terms: *selective agent* and *selective environment*.

SELECTIVE AGENTS AND SELECTIVE ENVIRONMENTS

Life-forms don't exist independently, of course. They're born into complex and ever-changing environments. Those environments include factors that can affect an individual's likelihood of passing its genes on to the next generation—its fitness. Such factors are often referred to as *selective agents*, and the whole slew of selective factors that pertain to any life-form can be considered its *selective environment*.

For example, in the Tai forest of southern Ivory Coast, African crowned eagles (*Stephanoaetus coronatus*)—spectacularly beautiful, brown, blue, and slate-gray raptors that can kill and transport animals weighing up to ten kilograms (twenty-two pounds)—have long been known to prey on small primates. A recent study of primate bones discovered on the ground immediately beneath crowned eagle nests revealed that the eagles prey on adults as well as juveniles of several monkey species; many of the skulls have distinctive punctures from the eagles' powerful talons. Not only that, but the eagles are estimated to pick off about 2 percent of the forest's colobus monkeys, 13 percent of the mangabey monkeys, and up to 16 percent of the pottos (super-cute primates with large eyes and hairless ears).[16] In the Tai forest, then, we can say that for some primate populations—and certainly for the poor pottos—the crowned eagle is an important selective agent that can affect any individual's likelihood of passing its genes on to the next generation, largely by dining on it. A potto might wake up with a high fitness score one morning, but the moment an eagle spots it, that fitness score plummets, and it will hit zero the moment the eagle hits its target.

We're now considering the interaction of life-forms (the eagle and the monkeys and the pottos), and this will lead us into great complexity in chapter 8. For the moment, consider another example of a selective agent, although one engaged in a more complex interaction of species.

Caterpillars that will turn into Silvery Blue butterflies (*Glaucopsyche lygdamus*) have a "honeydew gland" that exudes a sweet substance that attracts ants like the common black ant (*Formica fusca*). The ants don't just drink the secretion; they also, subsequently, defend the caterpillars from certain flies and wasps that would otherwise parasitize the caterpillars by laying eggs in them, which would hatch into larvae that would kill the "host" caterpillar. When biologists working with a Colorado population of Silvery Blue butterfly caterpillars prevented the ants from defending the caterpillars, around half of the caterpillars were parasitized. Allowing the ants to defend the caterpillars, though, resulted in only about 20 percent of the caterpillars being parasitized. Significant selective agents, then, in the life of the Silvery Blue butterfly include the ants, which increase the chances of survival (increasing fitness of any given defended caterpillar), and the wasps and the flies, which decrease the chances of survival.[17] Here we can say that there is *selective pressure* for caterpillars to secrete particularly sweet ant attractant.

Predators and parasites aren't the only kinds of selective agents. As we've just seen, selective agents can be beneficial to an individual. Much of the literature of biology has focused on the negative effects of selective agents, obscuring the fact that there are many species interactions that are beneficial to both life-forms; these are referred to as *symbiotic* interactions, and we'll see more of them in chapter 8.

Further adding to the complexity of evolution is the fact that selective agents aren't always living things; again, *anything* that affects one's fitness—the likelihood of passing on one's genes to the next generation (or even the number of offspring one has compared to one's fellows)—can be considered a selective agent.

Bighorn sheep (*Ovis canadensis*), for example, are well suited to the cold conditions in Alberta, Canada. A twenty-one-year study found that bighorns regularly survive the –40 degrees Celsius/–40 degrees Fahrenheit winter storms that dump feet of snow in the mountains. However, when birthing occurs the following spring, newborns are likely to die if the weather is unfavorable; for example, if precipitation is low and therefore vegetation (for grazing) is low.[18] For newborns, then, an important selective agent is the amount of spring rain. Note that this selective agent—rainfall—isn't a conscious actor, it isn't trying to drive bighorn evolution this way or that, and its characteristics (a lot of rain or very little) are determined not by any conscious actors but simply by the physics of the rainfall-generating water cycle as they happen in any given spring. The point here is that many selective agents are simply environmental variables that have no interest in the welfare (or otherwise) of any life-form; they lack even the consciousness to have any interest in the first place. There's another lesson from this example: while spring rainfall is an important selective agent for juveniles, adults are better able to cope with food stress, whatever the time of year. This shows us that a selective agent might be important only at one period of life, whereas at some other time, it might be trivial.

As I mentioned earlier (and as we can see in nature every day) life-forms don't exist in isolation; they constantly interact with other life-forms, in all kinds of ways including eating them (for example, bighorn sheep munching on clover or grass), being eaten by them (for example, pottos being snatched up by crowned eagles), protecting them (ants guarding caterpillars), parasitizing them (wasps laying their eggs in caterpillars), and many others. Figure 4-1 presents some very simplified portraits of ecosystems. On top (A), you see a rain forest with distinctive plant and animal life at each level: the canopy, the understory, and the forest floor. Imagine trying to map out all the interactions of these life-forms, such as the mammals, birds, amphibians, insects, and plants in each level . . . and what do you do with the animals, for instance, that move on occasion from one level to another? In (B) we see a pleasant

Fig. 4-1. Ecosystems

scene of an oak tree with its associated birds, insects, acorns. . . . That is a small community, but it could take a lot of work to understand the connections between these life-forms—and what about the microbial life you can't even see here? Finally, in (C), we see a polar bear on ice, as part of an Arctic sea ice ecosystem including shrimp, crabs, birds, vegetation, of course, and, again, all the microscopic life as well.

Now, consider any single life-form living in any ecosystem and how complex its selective environment really is. For example, a volcanic eruption might affect weather in a way that reduces the population of a species that pollinates a flower that is eaten by a life-form's main prey!

It's necessary for field biologists to try to outline a given species' most important selective agents, but when you consider the following, that is a tall order:

- Selective agents might themselves change through time.
- Selective agents might not be important throughout the life course.
- New selective agents might arise at any moment.

You can imagine the complexity here. Selective agents are many and complexly intertwined; if they are alive, they themselves are dealing with selective agents! Add the dimension of time, and the sometimes unpredictable nonconscious natural processes such as plate tectonics or the impact of a comet on the earth (talk about a selective agent!), and the mind reels—and what about all that microscopic life? Some microbes help animals digest their food; others are lethal. How do you "outline" a life-form's selective environment? The best we can do is to develop long-term understanding of different life-forms and their selective agents and environments.[19]

POPULATION THINKING AND MODES OF SELECTION

Life-forms, we know now, are at birth immersed in complex selective environments. Based on their variations, some are more fit than others. Selection is not an act but simply the fact that those with greater fitness will tend to pass their genes on more often by having more offspring. All clear and observable in the natural world.

To go further we need to begin to think about populations rather than individuals. All life-forms—even the occasionally solitary ones, like male orangutans—are members of populations of their own kind. How selection plays out for the individual is one thing, and how it affects entire populations is something else. There are several ways that selection "plays out" in real-world populations. These are called *modes* of selection. Many have been proposed, but three are almost universally recognized.

NEGATIVE SELECTION

When members of a population are born with variations that reduce fitness, they will tend to have less offspring. Generally speaking, over time the fitness-reducing variation will become rare in the population, or it may even disappear. This is called "selection against" the fitness-reducing variations, or negative selection.[20] Notice that there is no actual choice or "selection" going on here; there is nobody making a decision to eliminate this variation or that one; all that is happening is that variations that reduce fitness don't tend to get passed on. You can see how misleading the term *selection* can be, implying as it does some kind of selector or some kind of unity to the process of selection. Fitness-reducing variations are simply less commonly spread through the population, and they might eventually disappear.

For example, in western Colorado's Rocky Mountains there are populations of the perennial larkspur (*Delphinium nelsonii*), a beautiful mountain plant with rich blue petals. Not all the plants are blue, however; about 10 percent of any meadow's larkspurs are albino, with white rather than blue petals. A study led by University of California–Riverside biologist Nickolas Waser found that the albino larkspurs had the same shape as blue larkspurs and that they produced as many seeds as blue larkspurs. The albinos also produced as much quality nectar as the blue larkspurs. So why were albinos rare? Waser's observations revealed that the albinos were visited about 25 percent less often than blue larkspurs by hummingbird and bumblebee *pollinators* (animals that move male plant sperm to female plant egg cells). Why was that? The albinos were no different in shape from the blue larkspurs, so there was no physical reason preventing the bumblebees and hummingbirds from getting their nectar. And the albinos weren't genetically deficient in the number of seeds they produced. Something else was preventing pollinators from visiting them. Close examination revealed that in blue flowers, two of the petals nearest the nectaries (where pollinators suck up nectar) were actually white, forming a sort of easily visible "target" for the pollinators. In the albinos, though, while the nectaries were just fine and made plenty of good nectar, there was no such color difference, no "target" for the pollinators to spot. Consequently, pollinators spent more time and energy trying to get nectar from the albino plants, and therefore preferred the easier-to-drink-from "targeted" blue flowers.[21] Since petal color is genetically determined, this lower rate of pollination results in fewer larkspur seeds carrying DNA for albino coloration; this we can call "selection against" albinism, or "negative selection."

A natural question arises: why does the albino form persist? Why isn't it completely wiped out? One reason is that albinos are not entirely ignored; they are just visited less frequently than blue larkspurs. Also, albino coloration might be genetically linked to some other characteristic of the plant that simply can't be modified if the plant is to survive at all. In chapter 2, we saw that genes determine phenotypic (bodily) characteristics; but I haven't yet mentioned that some genes control multiple characteristics (such as albinism); this is called *pleitropy*. This may be the case with larkspur albinism. If albinism—which isn't fatal to the larkspur but does reduce its fitness somewhat—is controlled by a gene that simultaneously controls critical developmental processes (for example, early growth of the plant, which, if disturbed, *could* be fatal), the persistence of albinism might be explained.

POSITIVE SELECTION

As opposed to negative selection, when members of a population are born with variations that *increase* fitness, they will tend to have more offspring. Generally speaking, over time the fitness-increasing variation will become more common in the population. This is called "selection for" the fitness-increasing variation, or positive selection.[22] Notice that, once again, there is no actual choice or "selection" going on here; there is nobody making a decision to promote this variation or that one. All that is happening is that variations that increase fitness tend to be passed on and become more common in the population. And again, you can see how misleading the term *selection* can be because it implies some kind of selector, intent, or unity to the process of "selection." But there's no *there* there; fitness-increasing variations are simply passed on more often, becoming more common in the population; eventually they might become necessary for survival, such that every member of the population will have them.

Good examples of positive selection are easy to find in our own species. Although, for reasons I outlined in the introduction, I prefer not to focus on human evolution in this book, here I'd like to present a good example of positive selection in modern humanity.

This is the case of lactose tolerance. Most non-European people are lactose intolerant, meaning that by adulthood they can no longer digest lactose, a component of milk; almost 100 percent of Southeast Asian people, for example, cannot digest milk. But most Europeans are lactose tolerant. Lactose digestion requires the proper formation of *lactase*. Production of lactase is controlled by a gene called LCT. You can imagine what was done to understand the genetic basis of lactose intolerance. The LCT gene sequence (remember, from chapter 2 that a gene is a specific sequence of As, Cs, Gs and Ts) of nearly two hundred lactose-digesting Europeans and non-lactose-digesting non-Europeans were compared. The result: the LCT gene was slightly different (a mutation) in Europeans, producing a lactase that allows lactose digestion. Non-Europeans didn't have the mutation and therefore could not digest milk. The inference is clear; somewhere in the history of the European population, a mutation occurred in the LCT gene that allowed lactose digestion. Remember, nobody made a decision that people should suddenly be able to digest milk; it was simply a chance mutation that occurred and, for whatever reason (perhaps because milk provides vitamin D, energy, protein, and calcium), increased fitness by allowing the digestion of milk.[23] Since European populations are known to have started domesticating cows over five thousand years ago,[24] it's pretty clear that somewhere in these last five thousand years the mutation showed up, somewhere in Europe. Being a fitness-increasing variation, it proliferated; we can call this "positive selection."

But why don't non-Europeans have this mutation? Well, a few do because of the wide interbreeding of Europeans with people worldwide in the last few centuries, which spread the variation—the gene—for lactose tolerance. But there are also non-European populations who have a similar mutation—but not the *same* mutation—which also allows lactose digestion.[25] These people are all *pastoralists*, people who raise cows and drink their milk, including the Tutsi and Fulani people of Africa. The fact that the African variation is similar but not identical to the European variation strongly suggests that these mutations arose independently.[26]

BALANCING SELECTION

In *balancing selection*, more than one fitness-increasing variation proliferates simultaneously. Often, the variations that are being "selected for" are quite different. Generally speaking, over time these fitness-increasing variations will become more common in the population, and because of this the population might even diverge into two new populations, each characterized by this new variation.[27] Notice that, yet again, there is no actual choice or "selection" going on here; nobody makes a decision to promote these two variations. And again, you can see how misleading the term *selection* is, implying some kind of selector, intent, or unity in what's happening. But again, there's no *there* there—more than one fitness-increasing variation is simply spreading in the population because they each increase fitness, and fitness is a measure of the relationship of one's variations—one's essential form—to the properties of the selective environment.

One good example of balancing selection is found in Africa. There, the black-bellied seedcracker (*Pyrenestes ostrinus*), a common African finch, subsists largely on the seeds of two kinds of sedges; the bush knife (*Scleria verruscosa*), which grows up to three meters (ten feet) tall, and the nutrush (*Scleria goossensii*). As early as 1805, French naturalist Louis Jean Pierre Vieillot[28] (1748–1831) described seedcrackers, and later observers noted that some of these blood-red and black birds had beaks slightly larger than others. To verify and investigate this scientifically, between 1983 and 1990 biologist Thomas Bates Smith made an extensive study of seedcrackers in Cameroon. Netting and measuring nearly three thousand seedcrackers revealed that, indeed, the smaller beaks were around 13 millimeters (.51 inches) in width, while the larger beaks were around 15.5 millimeters (.61 inches) in width. That few millimeters (a tenth of an inch) might not seem like much, but if you're a seedcracker, the difference is significant because the seeds of the nutrush sedge are much softer than the seeds of the bush knife sedge. In this population, seedcrackers with smaller beaks were observed to prefer the softer seeds, while seedcrackers with larger beaks more commonly ate the harder seeds. A dietary difference, then, between very different food hardness was reflected in different beak sizes.

If seedcracker beak width were controlled by, say, the specific chemistry of the water or some other environmental factor, this wouldn't be a case of selection on variation in beak width, because selection acts, largely speaking, on genetically controlled characteristics. That is, for a characteristic to be selected for or against, it has to be encoded in the genes, because it is the genes that build the next generation. To be sure that beak size wasn't somehow controlled by some environmental variable, Smith had ninety-seven of the birds sent back to his lab in the United States, where he bred them for some time and found that beak size was genetically, and not environmentally, controlled; seedcrackers, even in their artificial environment, were born with beaks that quickly developed into only either the large or small form, rather than anywhere in between. Therefore, seedcrackers in the wild are born with beaks that are genetically destined to become large or small. The presence of these two significantly different beak sizes in the seedcracker population is called balancing selection.[29]

Another example of selection simultaneously favoring two quite different variations in the same population comes from the rainy Pacific Northwest Coast. Here, males of wild Coho salmon (*Oncorhynchus kisutch*) are normally either large-bodied (up to about 70 centimeters [27 inches]) or small-bodied (up to about 40 centimeters [15.7 inches]). Born in mainland freshwater streams, Coho salmon spend a year there as fry before migrating to the ocean as smolt. In the ocean, they might spend from half a year to two years before turning back toward the mainland and swimming upstream to their spawning grounds. By this time, while females average 50 centimeters (about 20 inches) in length, most males are either of the large form, called hooknose, or of the small form, called jacks. Up the streams they thrash. At the spawning ground, females excavate a nest, facing upstream and using their tails to swish away sand, creating a circular depression in the sandy streambed. Now the hooknose males (equipped with some pretty fearsome snaggle-teeth) compete intensely for the privilege of depositing their sperm on the eggs the female lays in the nest. Of the hooknose males, the best fighters are more likely to deposit their sperm, thus contributing to the next generation; they are more fit, then, than the slightly less-adept hooknose fighters. But is the giant hooknose more fit than the smaller jack? Not at all. Jacks also deposit sperm on the eggs in the nest, but they do so by darting out from hiding spaces—behind vegetation or rocks, for example—to deposit their sperm on the egg nest before flitting back under cover. They're in and out before the hooknose can react. And medium-sized Coho? There are few—being too small to fight the hooknose and too large to effectively hide. This, then, we can call balancing selection; two morphs, the large and the small, are "selected for."[30]

Balancing selection is not as well understood as positive and negative selection, and we should remember that "how evolution works" is a work in progress.[31]

SEXUAL SELECTION

While we've seen that intent is difficult to imagine in positive, negative, or balancing selection—because no real decisions were being made about any particular variation—there is one kind of selection in which real choices are made. This is called *sexual selection*, in which an organism does not mate with just any suitor but selects (sometimes with a variety of decision-making consciousness) among the various suitors. This, as we'll see below, can select for some characteristics and against some characteristics.

While many descriptions of sexual selection emphasize that males often compete for breeding access to females, there are plenty of cases—so far documented in birds, insects, fish, and mammals—in which males make mate-selection choices, leading to female competition for male access.[32]

Among the highly social meerkats (*Suricata suricatta*) and the South African mongooses we'll meet again in chapter 7, females compete vigorously for breeding opportunities with males, and this drives selection for larger body size in females (because larger body size helps a female dominate and defeat another female). In one study, females that achieved dominance in competitions weighed on average 699 grams (24.6 ounces) compared to the losers, who weighed on average 665 grams (23.4 ounces). As was the case with the seedcracker's beak size, this difference may not seem like much to you and me, but in the meerkat females it was found to be a statistically significant difference, meaning that it matters in the real world of meerkat life, not just in the human statistician's computations. If you're a fighting female meerkat, those few extra grams can be "the difference that makes a difference."[33]

Sexual selection can drive the spread of variations even more distinctive than overall body size; this is called the evolution of *secondary sexual characters* (*character* here meaning a trait or characteristic of the life-form).[34]

A good example is seen in the South American Hercules beetle (*Dynastes hercules*), in which males—but not females—possess gigantic pincers. In their leafy, South American forest-floor habitats, Herculean battles take place as males fight intensely for mating access to females. Explorer and naturalist William Beebe (1877–1962) described one battle in his dramatically titled 1947 research report "Notes on the Hercules Beetle *Dynastes hercules* (Linn.), at Rancho Grande, Venezuela, with Special Reference to Combat Behavior":

> The projecting horns touch and click, spread wide and close, the whole object of this opening phase being to get a grip on the opponent's horns. When the four horns are closed together, there is a deadlock.... Once this hold is attained and a firm grip secured the beetle rears up and up to an unbelievably vertical stance.... This posture is sustained for from two to as many as eight seconds, when the victim is either slammed down, or is carried away in some indefinite direction to some indeterminate distance, at the end of which the banging to earth will take place.[35]

For the male Hercules beetle, then, the "difference that makes a difference" here may well be the size and power of his fighting equipment. Put another way, his *fitness*—the likelihood that he will pass the genes for his own horns on to a new generation—is in part a measure of the size and effectiveness of those horns; therefore, there is sexual selection "for" more combat-effective horns. And, of course, the size and other characteristics of those horns are going to vary in any population; for this reason we have the elegant term "selection on variation."

But remember, not all is combat; like any life-form, the male Hercules has to find food, water, and nutrients, and his world is not one of simple fighting but a constant and multi-channeled negotiation of dozens of selective pressures right through from the moment his first cells form to the moment of death.

Sexual selection is a widely studied aspect of evolution, and—as with any other aspect of evolution—debate is lively about its precise workings and effects. Still, it is a significant factor of the evolution of many species. Biologist E. O. Wilson (b. 1929) has classified the major features of sexual selection, which I summarize below:

- **Choices made by males or females:** males or females choose a suitor from many who may or may not be in combative competition
- **Direct competition:** males or females engage in combat to gain access to the opposite sex, perhaps even establishing territories
- **Postcopulatory behavior:** "closing" of mating opportunity, for example, mated pair might physically move away from competing suitors[36]

Keep in mind that while mate choice superficially appears to be about some concern for the future, some concern for "the good of the species," there is little evidence that most life-forms—like female Hercules beetles, for example—are actually thinking ahead and selecting suitors that will send the "best genes" on to the next generation. Driven by instinct, it seems much more likely that the vast majority of sexual selection choices are driven by the fact that a suitor simply prevents another suitor from mating, or, if an active choice is made, that choice is about the suitor as it is today, right now; and the one that is today, right now, is most fit for the intricacies of finding a mate it is selected for mating. By this I mean that we have no data indicating that many life-forms make the kind of future-predicting choices many humans make, for example, in complex cultural marriage arrangements including specific rules for the division of wealth, inheritance, and so on. In most of the nonhuman animal world, suitors are chosen—selected for—because they possess characteristics that make them healthy enough to access a mate in the present moment.[37]

Table 4-1 summarizes the main modes of selection we've reviewed so far.

SELECTION: LESSONS SO FAR

What we've seen so far—in positive, negative, balancing, and sexual selection—is that the offspring of a given generation of some life-form are born into a "selective environment," and that environment "selects" for or against certain characteristics of that offspring, characteristics I generally refer to as variations. But remember, in many cases there is no real selection going on; there are simply the properties of the environment the life-form is born into, and its fit to that environment, which is based on what variations it possesses compared to its fellows. That is all. The word *selection* is misleading and, unfortunately, deeply ingrained. It implies a conscious "selector," but there is no evidence for one.

A final issue in selection is the various scales on which it occurs.

LEVELS OF SELECTION

A lot of debate in biology has centered on what level or levels selection acts. It's widely recognized that life is organized in a sort of hierarchy (imagine cells, organisms, populations, and so on), and it can be debated how different kinds of selection affect different levels of organization. Evolutionary biologist John A. Endler indicates that "[i]t is theoretically

Table 4-1. Modes of Selection

Kind of Selection	Factual Basis	Example
Negative	Variations that decrease fitness are less often passed on to the next generation.	A fly born without wings will be less capable of finding sustenance than its fellows, will live in poorer health than its fellows, and will find fewer mates than its fellows, so the genes for being born without wings will tend to decrease in the population.
Positive	Variations that increase fitness are more often passed on to the next generation.	An ibex born with particularly good balance will be more capable of surviving on its cliffside environment and is more likely to survive to reproductive age than its fellows, so the genes for better balance will tend to increase in the population.
Balancing	Variations that occur on either end of a spectrum sometimes *each* increase fitness, and so are more often passed on to the next generation than *mid-spectrum* variation.	Snails born with camouflage for either open ground or shady regions; individuals born without camouflage for one of these two very different habitats will be more likely to be picked off by birds, so their variation is less likely to continue in the population.
Sexual	Not all suitors for reproduction are equally successful in finding a mate.	Female swallows prefer to mate with males that possess longer tails than males possessing shorter tails, so genes for longer tails will proliferate in the population.
Kin	Sacrifices made by individuals benefit those closest to them biologically.	Macaques (monkeys) born with an inherent willingness to protect their siblings during predator attack will themselves eventually be protected, increasing their health, leading to mating and spread of willingness to protect siblings.

possible for selection to have different effects at levels other than the individual phenotype [body built by the genes]—for example, on genes, genotypes, groups, populations, species, or even self-replicators."[38]

Still, this is a little vague, and we need some clarification from a closer look at the levels of biological organization. Most biologists recognize at least four levels.

Replicators are the genes (and the DNA) themselves; these, as we know, carry the instructions for building future generations. So far I've referred to replicators as the DNA molecules themselves; the genes, or in sum for a species, the genome. For an individual, the replicators can be considered the genotype. Why they are considered separate from the things they build will be made clear below.

Interactors are the bodies (whether plant, animal, or what have you) built according to the information encoded in the replicator. So far I've called "interactors" *phenotypes*, the things built by the DNA. Remember that while interactors are close copies of their parents (because of the high fidelity of replicator copying across generations), they normally have variations, as we have seen.[39] Also recall that interactors are "born into selective environments."

So, the thing built—the body, or phenotype—is not the same thing as the genes that instruct building it. Why is this important? Because the phenotype does not last; it eventually dies, but the DNA moves on if that phenotype has managed to send its DNA on to the next generation by having offspring. As evolutionist Richard Dawkins pointed out in his mind-boggling book *The Selfish Gene*, a body—a phenotype—is just a gene's way of making another gene! Dawkins elaborates to say that even human identity is something of an illusion, that human beings are "survival machines" genetically programmed to propagate genes through time.[40] Many people are put off by Dawkins's coldly mechanistic terminology in that statement, but it's clear enough that the phenotype, the body, is indeed a biological structure that protects DNA and "pushes it into the future" by having offspring . . . food for thought!

In a similar way, G. C. Williams in his book *Natural Selection: Domains, Levels, and Challenges* distinguishes between two large domains of selection: the *codical* (the information-rich level of DNA) and the *material* (the things built by the DNA).[41]

The point regarding interactors is that they are very clearly subject to selection. One way to think of it is this: the natural world—the sum total of the selective pressures of one's selective environment—"evaluates" phenotypes. This is where fitness is computed. The results are selection. Everyone in biology agrees that interactors—the things built by DNA, the bodies, the organisms—are subject to natural selection.[42]

Kin-related groups of interactors are also real. No life-form is an island: a recent paper (which we'll come to in chapter 8) is titled "The Social Lives of Microbes"! All life-forms live in communities. Clearly, in sexually replicating species, many individuals live complex social lives in proximity to their immediate relatives or kin, like orangutans who care for their offspring for years before the offspring become independent. *Kin selection* is accepted by most biologists; it is the behavior of one individual to benefit another, closely related individual, at one's own expense. Although exactly how kin selection plays out (it's often called "altruism," in which X does something that will improve Y's fitness at the expense of X's fitness), most biologists accept that it occurs.

Local breeding groups are groups of kin-related groups themselves, sometimes known as *demes*; and *populations* are communities of individuals, of the same life-form (species), "linked by bonds of mating and parenthood; in other words, a population is a community of the same species."[43] While some evolutionists argue that selection can act on such large scales, others disagree; it's a fascinating debate, but not one I'm going to get into here. The point is, all populations are composed of individuals, and everyone agrees that selection acts on individuals.[44]

Species are essentially groups of life-forms that breed among themselves (we'll examine the species concept in detail in chapter 5). Does species-level selection exist? That is, are entire interbreeding populations, species, susceptible to the same bout of selection at the same time? There is plenty of debate about this.[45] The best case for it has been made by the prominent evolutionist Elizabeth Vrba, who proposes that the relatively simultaneous extinction of entire groups of life-forms—species—is clear in the fossil record of many African species, caused by the well-documented climate and vegetation changes of Africa around two to three million years ago. As large areas of forest gave way to grassland as a result of the beginning of the ice ages, Vrba argues, many forest-dwelling species gave way to grassland-dwelling species; such large-scale events are sometimes referred to as *turnovers*.[46]

Whatever biologists decide regarding the significance of selection on these various levels, or "units" of living things, keep in mind that it is all still selection; the fact that some offspring possess variations that make them more fit than others of their population. That, inevitably, leads to the propagation of some variations and the reduction of others; selection for, and selection against. As selective pressures change simply because conditions change, different variations make for fitness over time, and therefore the form of life-forms—what they look like, what they eat, what they do—changes over time.

Again, there has been, and still is, plenty of debate about exactly how selection works on these different levels of organization. Even in a textbook on evolutionary genetics, evolutionary biologist Francisco J. Ayala (b. 1934)—recipient of both the 2010 Templeton Prize and the 2002 National Science Medal—says of these levels: "The concept of local population may seem clear, but its application in practice entails difficulties because the boundaries between local populations are often fuzzy."[47]

Figure 4-2 shows the geographical distribution of the sand lizard (*Lacerta agilis*) in the shaded area (A). On the right is a box where we zoom in (lower panel), seeing that mountain ranges (B, C, and D) separate streams (thin gray lines) that flow into a river (E). Zooming in again (left panel) we see two gray areas (F and G), each of which encompasses a *local* population. Inside each local population area we see several *demes*, labeled (H). Each deme itself is composed of twenty or more black dots, and each dot itself is a *family group* consisting of around six individuals (I). A little math shows us that each deme is composed of about 120 individuals (the one on the lower right is particularly large), and each local population is composed of about 360 individuals.[48]

Again, while life scientists hold differing views about the levels on which selection really works, all would recognize the essential reality of these levels in the real world. By that I mean that if humans were to become extinct tomorrow, these boundaries (however fuzzy) wouldn't just dissolve; they are a reality external to human consciousness, and sand lizards of a breeding population in, say, northern Germany would not instantly begin breeding with members of their own species in eastern Russia. That just couldn't physically happen. There is reality in some of these levels, however fuzzy they are when we try to draw lines around them for analytical purposes.[49]

The one level of selection that everyone agrees on is that of the individual. Individual life-forms, close approximations of their parents (because of the quality of DNA replication) but carrying variations from those parents and their siblings (for a number of documented reasons), don't all have the same fitness; some are more likely to pass on the genes that built them to the next generation by having offspring. When an individual that has not had offspring dies, its death is called a *genetic*

Fig. 4-2. Sand Lizard Distribution

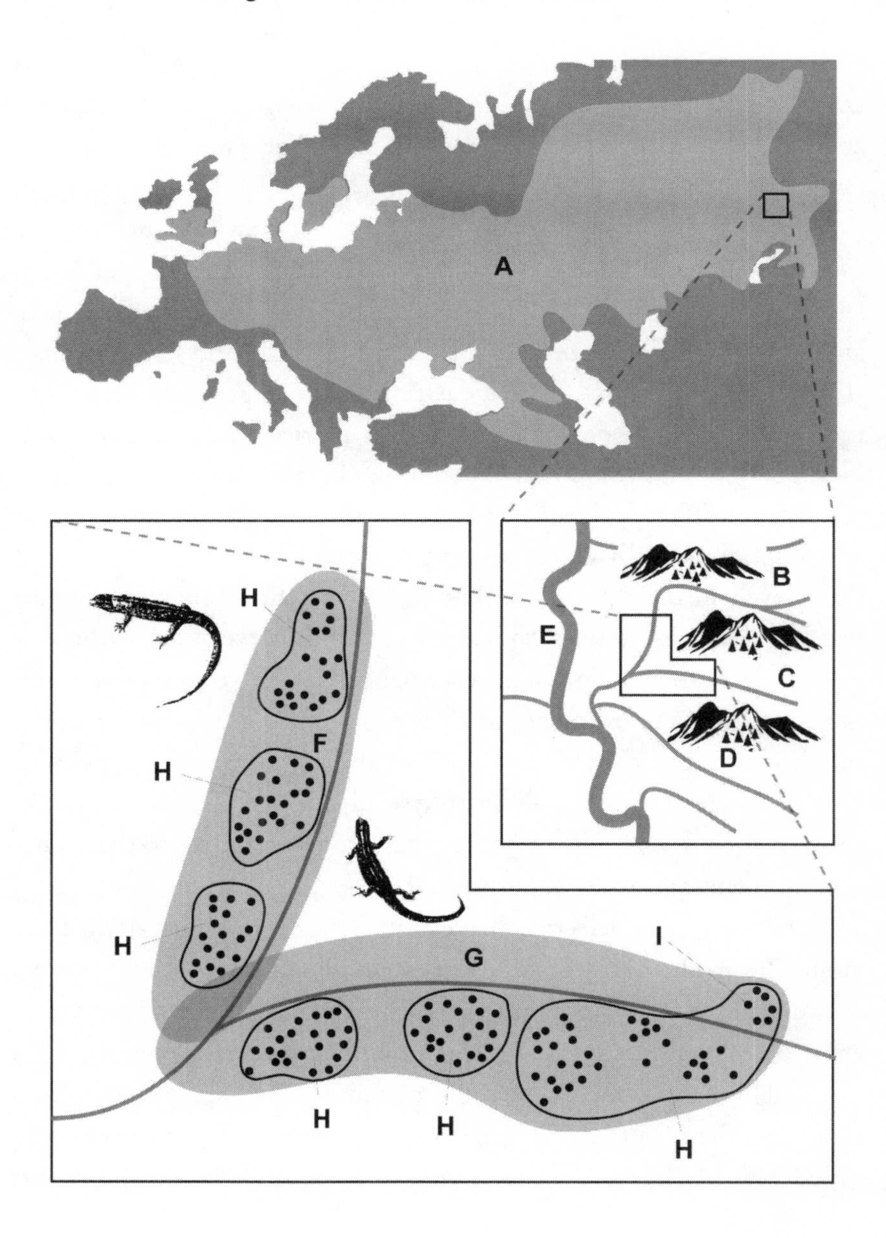

death. The genes that built that less-fit individual—that one individual of its entire population—are not passed on. This is selection against the individual—the interactor, or phenotype, as mentioned above. Since that individual carried genes that built it, we can say that selection has been "against" the genes that gave that individual a lower fitness than its fellows (for example, selection against the genes for building a fly that has no, or significantly deformed, wings).

In one important way, then, the real significance of the question of levels of life organization is that of selection itself: does selection act on genes, individuals, populations, species, or even larger units? At the least, everyone agrees that selection "sees" or "acts" on the phenotype, the thing built by the genotype. The evolutionist Stephen Jay Gould put it clearly: "Selection views bodies."[50] To clarify what a body is, biologist John A. Endler reminds us that "[a]n organism does not consist of a bag of traits, each of which can be considered in isolation. . . . Natural selection affects the whole organism, and many of an organism's traits will contribute to its ability to mate and survive."[51]

Table 4-2 summarizes levels of biological organization that most biologists would accept.[52]

To bring all this together visually, figure 4-3 shows three main levels of biological organization. In the upper panel, you see the DNA, the microscopic chemical structure that assembles the twenty amino acids into the thousands of proteins that build bodies. This is the realm of the genotype, the replicator. In the middle we see a variety of organisms built by DNA; this is the world of the phenotype, the interactor, the thing actually built by the DNA (the phenotype can include behavior—what the life-form *does*, but that is not easy to see here). In the bottom panel we see that individuals are not isolated but exist in complex selective environments, and in fact are organized in reproductive groups (for example, demes), larger populations (for example, the Canadian and

Table 4-2. Levels of Biological Organization

Theoretical Level	Real-World Manifestation	Example
Replicator	Genes = information-rich arrangements of matter	Starfish; humans
Interactor	Phenotypes = organisms built by genetic code	Thousands of starfish genes compose the starfish genome, and though this genome shares some information with the human genome, they are significantly different.
Actual Reproductive Group	Deme = local reproductive community (may be connected "down the line" to other demes)	Utah tortoises of the same species as Mexican tortoises live apart, but the two will probably never meet; their genes might meet, however, by "down the line" movement of genes through immediately adjacent local reproductive communities from Utah to Mexico.
Potential Genetic Group	Species = all reproductively compatible organisms of a single type	All members of the black rat species (*Rattus rattus*) worldwide.
Selective Environment	Ecology = all other life-forms and other selective pressures relevant to a given individual, deme, and/or species Communities (composed of interacting species) of communities are sometimes considered ecosystems.	Arctic tundra plants are eaten by lemmings; lemmings are eaten by Arctic foxes; Arctic foxes are eaten by wolves; wolves are hosts to parasites; parasites may be affected by temperature; humans hunt wolves; humans fall through sea ice if they have not learned how to identify weak spots; reindeer migrations are affected by annual snow cover; human reindeer hunting activities are affected by reindeer migrations.

Fig. 4-3. Major Levels of Living Things

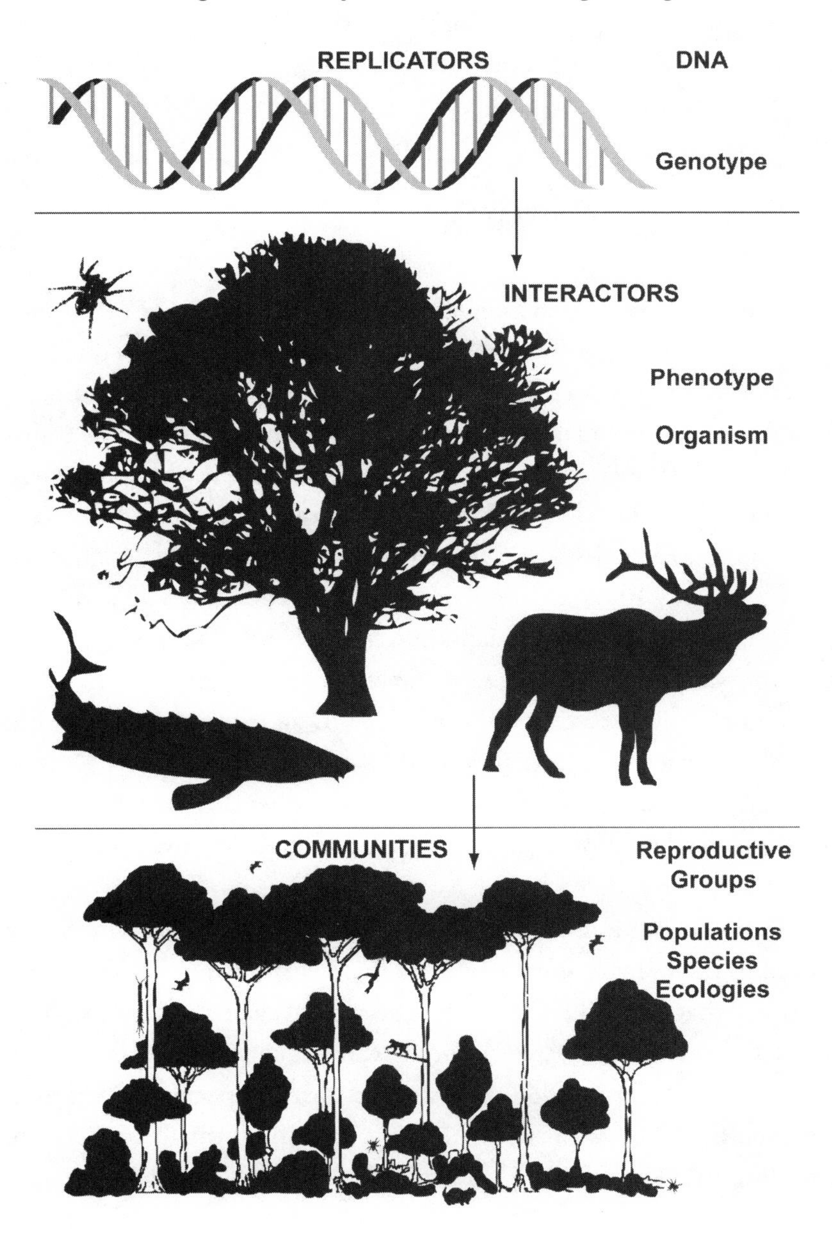

Alaskan polar bear populations, whose ranges might overlap), and larger groupings called species, which breed only with their own kind. Every life-form is a member of some kind of ecology, a web of connections among life-forms.

SELECTION AND ADAPTATION

I've avoided using the term *adaptation* so far, largely because of the baggage attached to it in the mass media. Adaptation is commonly linked to the idea of "evolutionary progress." As with other concepts, the usual feeling we get about adaptation from the mass media is that it is something that life-forms strive for, something that is somehow sought out. But while adaptation does occur, and it can carry the illusion of something sought as a goal—a beaver really does seem perfectly designed for its environment—the fact is that adaptation as a goal is an illusion.

To clarify, let me define adaptation in a biological sense: an adaptation is, to paraphrase biogeographer Geerat J. Vermeij, any variation that allows an organism to live and reproduce in an environment where it probably otherwise couldn't survive.[53]

A little more narrowly, and closer to the level of the individual organism, an adaptation can be considered something (a variation) that improves fitness.

Clearly, adaptations do exist, and species do over time become better tailored to their environments because there will always be selection for variations that improve fitness and selection against variations that reduce fitness. In this way, evolution can be considered progressive; progress and adaptation do occur as a life-form (a species) is "tailored" over time by selective pressures, "fitting" the life-form to its environment. But notice that I've had to put "tailored" (and "fitting") into quotation marks; the word *tailor* implies some entity that makes things for a certain purpose, like a person who designs and makes business suits or track shoes.

But in the natural world, we don't see this. In no species other than humanity do we see a conscious effort to drive the variations of a given life-form one way or another. Humans certainly do select certain variations in the plants and animals we farm, and we certainly do propagate them at the expense of others, tailoring (without quotation marks) our farmed life-forms for our wishes. But again, this is absent in nature. Selective environments, we've seen, are composed of many selective agents, many of which are simply physical properties of an environment (like temperature) that have no apparent consciousness with which to tailor anything. So adaptation does happen, and progress does happen, and species do over time become better adapted to their environments because fitness-increasing variations tend to be passed on and fitness-decreasing variations tend to be lost. That is evolution.

SELECTION IN EVOLUTION

We are back where we started, with selection on variation. Despite the intent implied in the word *selection*, selection is not a thing or a conscious agent. Selection simply refers to the fact that not all members of a population have the same number of offspring; many die before reaching reproductive age, and those that do vary in how many offspring they produce. All this is because life-forms aren't clones; they have variations that make some fitter than their fellows. We can observe these fitness differences in the natural world, all around us, every day, just as we can observe the fact that not all life-forms "push" their DNA into the future by having offspring.

Which variations are going to come up is unplanned and essentially random (though constraints are important), but which members of a population pass on their genes will not be random; the fitter—those better adapted to their selective environment—will pass on their genes more often than the less fit. Over time, the characteristics of the population will change because the environment will change, meaning that "what it is to be fit" will change.

Selection is not a concrete "thing" but an observable fact. As it occurs, the characteristics of life-forms change through time; sometimes they change enough that a new kind of life emerges from a previous kind of life; this is speciation, the subject of the next chapter.

CHAPTER 5
THE FACT OF SPECIATION

> **First, then, if things were made out of nothing, any species could spring from any source and nothing would require seed. Men could arise from the sea and scaly fish from the earth, and birds could be hatched out of the sky. . . . The same fruits would not grow constantly on the same trees, but they would keep changing; any tree might bear any fruit. If each species were not composed of its own generative bodies why should each be born always of the same kind of mother? Actually, since each is formed out of specific seeds, it is born and emerges into the sunlit world only from a place where there exists the right material, the right kind of atoms. This is why everything cannot be born of everything, but a specific power of generation inheres in specific objects.**
>
> —*Lucretius: On the Nature of the Universe*[1]

So we see that replication is a fact; offspring come from parental generations. And variation is a fact: offspring aren't normally clones; they differ from their parents and among their siblings. Selection is also a fact: of the variable offspring, those with the best fit to their selective environment will tend to pass on the genes that made them more often than those with a poorer fit. These three facts cannot be rationally disputed.

Based on some definitions of evolution, that's all we need to say. If the genes in a population are changing, as far as population geneticists are concerned, then the organisms built by these genes are changing, and evolution is happening. That's because of a far-ranging and important implication, which is this: give this system of replication, variation, and selection long enough, alter the selective pressures a little here and there (as must occur because the universe changes; it is not static), and you are going to get speciation—the appearance of new forms of life.

This chapter is about exactly how that occurs, and why we know that it does.

WHAT IS A SPECIES?

Speciation is what we most commonly associate with the word *evolution*; the appearance of new kind of life. We know new life-forms don't just pop up out of nowhere. To understand speciation, though, the appearance of new life-forms, we need to understand species.

The word *species*, the *Oxford English Dictionary* tells us, is derived from Latin and generally refers to "appearance" or "outward form."[2] In 1559, the term was used to differentiate between different kinds of wine. At the time, many new species were being discovered as explorers brought back to Europe tales of exotic beasts, and in *Antony and Cleopatra*, William Shakespeare (1564–1616) poked fun at some attempts to describe them:

> LEPIDUS: What manner o' thing is your crocodile?
> ANTONY: It is shaped, sir, like itself; and it is as broad as it hath

> breadth: it is just so high as it is, and moves with its own organs: it lives by that which nourisheth it, and the elements once out of it, it transmigrates.
>
> LEPIDUS: What color is it of?
>
> ANTONY: Of its own color too.
>
> LEPIDUS: 'Tis a strange serpent.
>
> ANTONY: 'Tis so. And the tears of it are wet.
>
> —William Shakespeare, *Antony and Cleopatra*, act 2, scene 7

In relation to the life sciences, *species* was first used in Edward Topsell's thrillingly illustrated tome *The Historie of Serpents* (1608),[3] and by 1700 the term was used freely to refer to obvious groupings of plants and animals; the zebra and the hawk were both clearly living things, as were oaks, but each was obviously a different kind of life, a different species. This was straightforward enough in those days of *descriptive* biology, when life-forms were largely thought to be unchanging *kinds* created by God. After all, it was written in Genesis that "God made the beast of the earth after his kind, and cattle after their kind, and every thing that creepeth upon the earth after his kind: and God saw that it was good" (Genesis 1:25).[4]

So there was not much to explain in this earlier period of Western science; the naturalist's function was to describe rather than to explain. Swedish naturalist Carolus Linnaeus (1707–1778) was the first to use the term systematically and wholesale in his multivolume description of the species of life known to him, the *Systema Naturae*.[5]

These descriptive roots of the Western study of life shaped the character of biology even into the 1920s. Although biologists knew that evolution was happening, their concept of a life-form, a species, was still somewhat fixed, the legacy of centuries of description. But in the 1930s, Ernst Mayr—the biologist we've met several times already—was just back from field studies in New Guinea and was pushing a new perspective: get off the concretized, pigeonhole mentality, he urged, and start thinking not just of the visible characteristics of a life-form but of even its behavior, its geographical distribution, and its entire population.

Individuals, Mayr argued, were of course important, but we don't see individuals evolve instantly into a new kind of life; what we see is that populations change as beneficial variations spread or as fitness-reducing variations are selected against.[6] In either case, it's populations that really change through time—resulting in new kinds of life—and therefore populations needed to be understood. Biology listened, and this shift toward population thinking is widely considered to be one of Mayr's fundamental contributions to the field.[7]

Another contribution—which we're more concerned with here—was Mayr's definition of species. Individuals are members of populations, and these populations, Mayr argued in 1942, were characterized largely by the fact that they bred within themselves and not with other populations. Species, Mayr stated, were "groups of actually or potentially interbreeding natural populations, which are reproductively isolated from other such groups."[8]

A couple of clarifications. First, by "actually or potentially," Mayr was pointing out that populations of the same species, the same kind of life, might be widespread in the world and yet not actually breeding with one another, although they could if they met. For example, North American tortoises (*Gopherus agassizii*) occupy a range from southern Utah to western Mexico, but the Utah and Mexico turtles will probably never meet because their home ranges do not overlap. Second, Mayr refers to "reproductive isolation," which is the real heart of the issue; species do not mate with members of other species; they are reproductively isolated. Bats do not attempt to mate with snails. Ostriches do not attempt to mate with sea slugs. Populations of clearly different life-forms are reproductively isolated.[9]

The Fact of Speciation

Although so much discussion of the species concept has taken place that the *Oxford English Dictionary* notes that it has been the subject of "much discussion,"[10] a broad range of recent reviews of Mayr's definition (widely known as the *biological species definition*) show that it is widely accepted.[11] The main reason that Mayr's concept has endured is that it is accurate. Fifty-three years after his 1942 publication, Mayr wrote a paper titled (showing a little frustration) "What Is a Species and What Is Not?" In the paper he reiterates that despite what he considered excessive philosophizing by "armchair" biologists, species were real in the same way that planets are real. Planets are exterior to humanity, they have been discovered rather than invented, and they would not—like the concept of ice cream, for example (an entirely human fabrication)—vanish if humanity were to vanish. In the same way, nothing in science tells us that life-forms would begin to attempt to breed with other life-forms, willy-nilly, if humans were to suddenly vanish. Species are real, they are exterior to humanity, and they are discoveries rather than inventions.[12]

Biological classification today includes three main levels of organization: genus, species, and (sometimes) subspecies. Some common species and their scientific (Latin) names (in italics), with notes regarding naming are shown in Table 5-1.

We'll come back to the species concept in chapter 8, but for the moment, it's enough to recognize that they are real. A recent review indicates that there are at least four very good criteria for identifying a species:

1. Members live in reproductive isolation.
2. Members share a mate-recognition system.
3. Members occupy the same ecological niche.
4. Members share a common ancestor.[13]

Table 5-1. Genus and Species Names of Some Life-Forms

Genus	Species (and Subspecies If Apparent)	Common Name and Comment
Homo	*sapiens sapiens*	Human (*Homo* refers to humanity; *sapiens* refers to *sapere* = to be wise)
Homo	*sapiens neanderthalensis*	Neanderthal (extinct human; first fossil remains found in Neander Valley, Germany)
Aptostichus	*stephencolberti*	Spider named after comedian Stephen Colbert in 2008
Pan	*troglodytes*	Chimpanzee (*troglo* refers to caves and to darkness, referring to the dark forests of Central Africa)
Felis	*domesticus*	Domesticated cat (*domesticus* refers to household)
Felis	*concolor*	Puma (*concolor* refers to its several coat colors)
Canis	*lupus familiaris*	Domesticated dog (see below for *lupus*)
Canis	*lupus*	Wolf (*lupus* is from Latin for "wolf," and it relates to the Spanish *lobo* = wolf)
Musca	*domestica*	House fly (*domestica* refers to household)
Helix	*aspersa*	Garden snail (*helix* refers to spiral)
Rosa	*woodsii*	Woods' rose (named for botanist Joseph Woods, 1776–1864)
Platanista	*gangetica gangetica*	Ganges river dolphin
Ruminococcus	*albus*	Bacterium that inhabits the cow rumen (stomach) and facilitates digestion of grass, producing milk

Finally, Ernst Mayr weighed in as recently as 2005, stating that his species concept was reasonable and remained up-to-date: "The actual demarcation of species taxa uses morphological, ecological, behavioral, and molecular information to infer the rank [species naming] of isolated populations."[14]

In the 1700s, Linnaeus described about seven thousand plant and animal species, and today we know of about five million; but some have suggested that this estimate may be ten times too low.[15] New species continue to be discovered—many in the oceans—but at the same time, humanity's activities around the globe are estimated by some to cause the extinction of up to three species per hour.[16]

Before we look at how these species originate—where they all came from, like the diversity of life-forms on the front cover of this book—I'd like to remind you of two points. First, recall that, as Mayr pointed out, species are not fixed, and they can be hard to "draw a line around." Some don't precisely fit the definition, and that means there are things we don't understand about biology. That's OK; rather than ditching biology, which has explained so much else, we just keep improving our knowledge. To remind us not to think typologically, even before Mayr, Theodosius Dobzhansky, referring to the passage of time, wrote that "[s]pecies is a stage in a process, not a static unit."[17]

Second, keep in mind that species aren't an *end product*; nothing known to science suggests that species *were made to do* anything. This lack of central control or decision making is hard to imagine, but it's important. The "perfectly functioning" ecosystems we see on nature shows are often an illusion; look at a species, and you might well see a life-form that's in trouble, becoming extinct before your eyes, but slowly. And even in the smoothly co-evolving ecosystems that we can say are at "equilibrium," remember that while the species do indeed co-evolve, the idea that they were made to co-evolve, that they have a certain function (something else TV writers often tell us), is simply wrong. In a review of *symbiosis*, the co-evolution of life-forms, microbiologist Lynn Margulis notes that she will not use the terms *competition* or *cooperation*: "These words may be appropriate for the basketball court, computer industry, and financial institutions, but they paint with too broad a brush."[18]

Life-forms do fall into mutual arrangements, and they do co-evolve, but they weren't tailored for those functions by some "controller," and they're not end products, perfectly shaped for a "job."[19] We'll revisit this issue in chapter 8.

We now have an idea of what species are: real populations of life-forms that can or do breed among themselves and are reproductively isolated from other living populations. Let's see where they come from.

WHAT IS SPECIATION?

Speciation is "the formation of two or more new species from one ancestral population."[20] Since a species is a population of life-forms that could or do mate within themselves, they are reproductively isolated from other living populations. How reproductive isolation happens is relatively straightforward, and it leads us right into the heart of speciation. There are two main kinds of reproductive isolation, and they're both normally studied by the examination of "sister," or very closely related, species, for reasons that will become clear in the next pages.

PREMATING REPRODUCTIVE ISOLATION

Premating isolating mechanisms prevent potential mates from encountering one another. This prevents the flow of genes between two populations of the same kind of life—the same species—and genetic cutoff or isolation is the key to speciation.

Premating isolation can occur when members of a population live out their lives, including the mating phase or phases of life, in different habitats. This is known as *ecological isolation*. It can be as dramatic as the case of the Utah and Mexico subpopulations of the desert turtle mentioned earlier; since these turtles simply don't migrate from Mexico to Utah (or vice versa), they might become reproductively isolated. Ecological isolation can also happen if some members of a single species spread into a relatively nearby habitat and remain there, as in the case of the North American threespine stickleback (*Gasterosteus aculeatus*). Some of these little (up to about 11 centimeters, or 4.3 inches) fishes spend most of their time feeding nearer the surface (*limnetics*), while others stay nearer the lake floor (*benthics*). Sunlight filters differently through shallower and deeper water, such that these two habitats have different light qualities. Over time, the females of the deeper-dwelling and shallower-dwelling groups have come to prefer to mate with males from their own deeper or shallower populations, leading to reproductive isolation.[21]

Behavioral isolation (also known as ethological isolation because *ethology* is the study of behavior) occurs in animals when subpopulations of a species develop behavioral differences that limit mating. An important component here is something we've seen before: sexual selection, or the preference of one sex for specific mates among many potential suitors. For example, in Africa's Lake Victoria, two genetically closely related species of cichlid (pronounced *sick-lid*) fishes, the excitable ruti (*Pundamilia nyererei*) and the zebra (*Pundamilia pundamilis*), differ mostly in their color patterns; both have vertical, stripy color patterns, but the rutis have yellow-orange coloration, whereas the zebra's colors range more toward blues. In laboratory experiments, females of each species strongly preferred to mate with males bearing color patterns of their own species. But when the aquarium lights were changed to prevent the females from distinguishing the males' color patterns, females no longer preferred to mate with members of their own species' color patterns. The inference, strongly supported by genetic data, is that a behavioral isolation developed in the wild, leading to speciation of these closely related species from an ancestral species.[22]

Another case comes in the form of North American ducks, the mallard (*Anas platyrhyncos*) and the pintail (*Anas acuta*). The males of these species are very similar in coloration, but female mallards have coloration very different from female pintails (and the pintail has a long, thin tail when viewed from the side). In the wild, pintail males mate with pintail females, and mallard males mate with mallard females. As closely related as mallards and pintails are—in the lab, mallards and pintails will mate—they do not mate in the wild. Why not? Because when they are mated, the offspring inherit quite different genes directing their mating displays; mallard–pintail hybrids don't act out their courtship displays in a way that attracts many mates, and therefore mallard–pintail hybrids in nature are "selected against," reinforcing mallard and pintail behavioral and reproductive isolation.[23]

A more intuitively direct kind of isolation is the differentiation, over time, of the reproductive organs themselves. The most compelling case comes from the two (once again) closely related wood beetle species of Japan (*Carabus maiyasanus* and *Carabus iwakianus*). The male and female genitalia of these species—as closely related as they are otherwise—are mismatched; copulations between the male of one species and the female of the other result in the males' genital structures being broken off and/or the females' genitals being severely damaged; in each case, successful transfer of sperm to egg is interrupted—the critical genetic isolation.[24] Mechanical isolation of this kind is often referred to an incompatibility of "lock and key."

A less well studied (but still viable) kind of premating isolation can occur when the mating schedules of subpopulations of a species become separated; this is *temporal* (*temporal* relating to time) *isolation*. For example, the closely related boulder star coral (*Montastrea annularis*) and the star coral (*Montastrea franksi*) differ mainly in the fact that each releases its sperm and eggs (in a spectacular mass release of thousands of particles, like soda fizz) about one and a half to three hours apart. This may not seem like a long time, but in the ocean environment the sperm quickly become crippled and unable to fertilize the egg cells. Nobody knows when or how the mating schedules diverged; one study found that these corals' egg and sperm releases could be experimentally altered by artificially simulating the sunset. But for whatever reason, the safe inference is that these two closely related species diverged when their mating schedules diverged.[25]

Another case of temporal reproductive isolation comes from two closely related species of coastal California pine tree, the Bishop (*Pinus muricata*) and the Monterey (*Pinus radiata*). The Monterey pine anthers (which release sperm) become active several months before the Bishop pine stigmas (which contain the egg) become active, so they cannot mate. It's again suspected that this led to a divergence into two species of pine where once there was only the one.[26]

Another case is seen in pink salmon (*Oncorhyncus gorbuscha*), whose members—again for reasons not well understood—can breed only in alternate years due to differences in their maturation process. After moving out to sea, they spend exactly two years (plus or minus only ten days, according to one study of Alaskan and western Canadian populations!) before they turn and migrate up the streams to the grounds where they were born. But a temporal gap comes up because the ones that were born in, let's say, 2000 came back to mate in 2002, whereas those born in 2001 came back to mate in 2003. This means that when one part of the population is ready to breed, the other is not. These salmon are today of the same species but in time may diverge into different species because of this kind of reproductive isolation.[27]

POSTMATING REPRODUCTIVE ISOLATION

Postmating isolating mechanisms prevent the successful development of an offspring when mating does occur between individuals of the same population of a species, despite the fact that their lives—their behavior, ecology, or reproductive apparatus—are quite different.

The best-known postmating isolation mechanism is *zygomatic mortality*, or the failure of the egg fertilized by the sperm (the zygote) to develop properly. This occurs because the DNA of the parental generations are simply too different—they have diverged over time so much—that they cannot properly join to build the offspring. Most other postmating reproduction-isolating mechanisms are the result of this same essential incompatibility of the male and female DNA.[28]

A few other forms of postmating isolation exist, including *hybrid inviability* (offspring form but die before achieving reproductive age) and *hybrid sterility* (offspring form, but—as in the case of a horse crossed with a donkey producing a sterile mule—their offspring do not have their own offspring).

Really, these terms, for our purposes, hardly matter. The point is that there are simple and well-known reasons for the appearance of basic incompatibilities of the male and female DNA, preventing successful reproduction. Just as effectively as geographical isolation, this kind of isolation sets in the thin end of the wedge that diverges populations into different forms: speciation.

The lesson from all these cases is that genetic cutoff is the key to the

Table 5-2. Major Reasons for Reproductive Isolation

Isolating Mechanism	**Kind of Isolation**	**Explanation**	**Example**
PREMATING	**Temporal Isolation**	Different mating schedules emerge.	American field crickets reach reproductive age in different seasons, preventing them from meeting and mating.
	Ecological Isolation	Different habitats are occupied.	Some Japanese ladybugs will mate only on the plant they prefer to eat; however, members of the same species eat different plants.
	Behavioral Isolation	Different behaviors become established.	Moths will attempt to mate only with females releasing a specific pheromone; fiddler crab females prefer males with larger claws.
	Mechanical Isolation	Male and female sexual organs are physically incompatible.	Closely related fruit fly species' male and female genitals simply do not fit as "lock and key."
POSTMATING	**Hybrid Inviability**	Hybrids of diverging populations do not develop properly.	Crosses of sheep and goats do not survive to reproductive age.
	Hybrid Sterility	Hybrids of diverging populations cannot have their own offspring.	Crosses of horses and donkeys are sterile.

divergence of life-forms. Table 5-2 summarizes isolating mechanisms for quick reference.

Reproductive isolation—genetic cutoff—divides populations, leading to speciation itself over time because the selective environments they now occupy will shape the newly separated groups in different ways.

There are three main ways that speciation plays out; you'll see how they involve isolating mechanisms as we explore them below. Since we know that the essence of reproductive isolation is the cutoff of genetic flow between groups, I'll refer to it from here on as genetic isolation.

KINDS OF SPECIATION

Because of the interest in speciation, a great variety of researchers have studied it for many years, and though they've learned a lot of important things, they've also accumulated a confusing grab bag of terms for the various ways that speciation can occur. I'll let Dutch biologist Menno Schilthuizen describe the current state of affairs: "A perusal of the evolutionary literature reveals a cornucopia of terms and theories, including microgeographic, semigeographic, single-gene, selfish-gene, symbiont, centrifugal, reinforcement, vicariant, competitive, dumb-bell, hybrid, ecological-adaptive, and—believe it or not—Adam and Eve speciation."[29]

To this list I can add allopatric, parapatric, peripatric, sympatric, stasi-parapatric, and alloparapatric. We will not be reviewing all these kinds of speciation because, at the least for our purposes and perhaps overall, as Schilthuizen continues: "Many are only different words for the same thing." Evolutionary biologist D. J. Futuyama agrees, stating that the kinds of speciation can be broadly classified by whether genetic isolation happens because of forces external to the population of a life-form or changes within that population.[30]

Forces *external* to the life-form include such large-scale changes as the joining of South America to Central America several million years ago (forming the land bridge we now call Panama); this effectively isolated the aquatic species of the Pacific and Atlantic Oceans. This is referred to as *allopatric speciation*, but here we can call it *external cutoff*. Changes *within* a population that lead to genetic isolation include the case of subpopulations of a life-form becoming so specialized to a certain resource that they cease interbreeding with the rest of the larger population; this is called *sympatric speciation*, but here we can call it *internal cutoff*. We'll start with external cutoff (or allopatric speciation).

EXTERNAL CUTOFF (ALLOPATRIC SPECIATION)

External cutoff (*allopatric*) speciation occurs when some group of a given population is geographically cut off from the rest, often by some geographical event. Examples are the best illustrations of the point.

One of the best-known examples of external cutoff speciation—total genetic isolation—is seen in the case I referred to above, that of the Atlantic and Pacific Oceans being separated by the isthmus of Panama (of course, the Atlantic and Pacific still connect at the southern tip of South America, so this applies only to the warmer-water species a bit north of the equator). Geological and other evidence has determined that the land bridge separating these bodies of waters came up between 3.0 and 3.5 million years ago. In the early 1990s, a Panamanian-American research team focused on the snapping shrimps of the Pacific and Atlantic (Caribbean) sides of Panama, focusing on their genetic similarities and differences and on the ability of the Caribbean and Pacific species to mate. Not only did the species differ in their coloration and other physical characteristics, but in matings between the Caribbean and Pacific forms, 99 percent of the offspring were unsuccessful. However, genetic data showed that these species were closely linked. And the genetic clock—a method that measures time by the steady accumulation of mutations in a genome over time—showed that these species diverged, or became genetically isolated, about three million years ago, corroborating the geological date of the separation of the Pacific and Caribbean.[31] All in all, this is a strong case for external cutoff (allopatric) speciation, in which a population that once freely interbred (revealed by great genetic similarity) became genetically isolated and then diverged into life-forms that can no longer significantly interbreed: speciation, the appearance of a new form of life.[32]

In the example we've just seen, substantial populations of the snapping shrimp were separated. In other cases, much smaller populations become isolated and diverge from their original populations. This often occurs when islands are colonized by just a small group of, say, a mainland population, either by migration or accident (for example, members of a continental species are carried by rafts of vegetation to a previously unpopulated island). Many cases are seen in the Hawaiian Islands, where many species of plants, snails, and flies are unique to the islands. In the case of the flies, the island of Hawaii has twenty-six species that are unique to it; however, those species are genetically most similar to those of the other islands, which came up from the sea *much earlier* than the island of Hawaii. The inference is that some populations of the older-island flies occasionally crossed the forty-six-kilometer (twenty-eight-mile) channel between the islands, diverging into new species on Hawaii.[33] Later in this chapter we'll see more instances of founder-effect speciation.

Another classic case of external cutoff is that of the many species of the Galapagos Islands. Among the finches (small birds of the genus *Geospiza*), thirteen species are known, and, once again, genetic analysis indicates that they are all closely related. Not only that, but the closest relatives of the Galapagos finches are the finches of mainland South America, nearly one thousand kilometers (over five hundred miles) away. The Galapagos finches have been clearly shaped by their selective environments; their beaks differ by species, and long-term studies have shown that some are most effective in crushing seeds, some in plucking flowers from cactus, some in snapping up insects, and so on. Each of these species "makes its living" in a slightly different way; not only that, but they have different songs. Yet, essentially, they are the same kind of bird—a finch, and that molecular evidence pointing to South America cannot be ignored. Since there is no evidence that finches arose independently on the Galapagos in the last five million years (when the islands formed), the Galapagos finches must have come from South America. Long-term study by biologists Peter and Rosemary Grant—who have visited the island almost every year since 1978 and are still going strong—has confirmed all these facts, indicating that since colonization of the Galapagos about 2.3 million years ago,[34] the finches have diversified into these many forms from the ancestral form.[35]

While the differences between large-population and small-population origins of new species are interesting, they're not something I want to worry about too much in this book.[36] The point is that genetic (reproductive) isolation—however it happens—leads, over time, to speciation.

INTERNAL CUTOFF (SYMPATRIC SPECIATION)

We started with the observation that populations can be split by geographical factors *external* to the population, and that's clear enough from the examples we've seen. There are also cases, though, where genetic isolation develops *within* a population of life-forms, and eventually the divergence—which begins with strong but not complete genetic isolation—leads to new species. This is *sympatric speciation*, but, as I mentioned, here I prefer to call it *internal cutoff*.

For example, the greenish warblers (genus *Phylloscopus*) of Asia are widely spread in a great ring around the Tibetan plateau, a geographical feature that averages about 4,500 meters in altitude (about 14,700 feet—a bit higher than the summit of Mount Rainier). Several studies of this species have shown that the warblers to the north, south, west, and east of the plateau are all genetically similar enough to interbreed—fulfilling the criteria of a single species—but also that they do not all interbreed. That's not because they're completely cut off; populations adjacent to one another all around the plateau *could* interbreed. But the four main groups around the plateau (which the birds don't fly over) have developed, over time, distinctive songs that seem to prevent interbreeding because of behavioral preferences—genetic isolation that has led to differentiation (speciation) around the Tibetan plateau.[37]

There are other kinds of speciation, as I mentioned before, but they're beyond the scope of this book. My point has been to show that the essential factor is genetic cutoff, of whatever kind. Now, cutoff alone doesn't necessarily lead to speciation, but since environments change, selective environments change, and when populations are cut off they are probably going to be reshaped, over time, by the selective environments of their new habitats, leading to enough divergence that speciation occurs.

Lesson: reproductive, genetic cutoff leads to isolation, which leads to accumulation of differences between populations that were once one, which leads to speciation.

Table 5-3 summarizes the main modes of speciation for quick reference.

Table 5-3. Modes of Speciation

Mode of Speciation	Why Populations Become Isolated	Why Populations Diverge	Example
External Cutoff (Allopatric Speciation)	Total geographical separation of previously interbreeding populations	Populations in regions (A) and (B) adapt to those regions over time, accumulating differences.	Land bridge forming Panama separates Pacific and Atlantic shrimp populations, which subsequently diverge.
Internal Cutoff (Sympatric Speciation)	Behavioral isolation of previously interbreeding populations	Multiple forms of the original population develop by sexual or balancing selection or specialization, for example, on a limited food source.	Mating preferences in some African lake fish diverged over time, leading to reproductive isolation and genetic divergence.

OBSERVING SPECIATION

In 1922, the early geneticist William Bateson (1861–1926) wrote, "In dim outline evolution is evident enough. But that particular and essential bit of the theory of evolution which is concerned with the origin and nature of species remains utterly mysterious."[38]

Things have changed a lot, as you're about to see.

What critics of evolution often want to see as proof of evolution is speciation itself, the appearance of a new life-form from an older life-form. The absence of easily observable speciation is, to many, evidence that evolution "is wrong." We'll see that evolution isn't wrong, but let's start with three reasons that speciation can be tricky to observe.

First, speciation generally occurs slowly, normally over centuries at least. After reproductive isolation happens, enough new variations have to accumulate that the once-united populations could no longer interbreed. But even then, when do you draw the line? Do you do experiments with every single potential mating pair of a population? Do you consider it speciation only when no matings produce healthy offspring, or is 70 percent failure sufficient to "make the call"? It is impractical to test speciation in this way, just as it was impractical to test certain aspects of gravitation used in the space program; but time and again, using mathematics to make predictions, our lunar landers and other spacecraft have gone exactly where theory said they should go. We would be foolish to argue that the back side of the moon simply didn't exist until human eyes observed it directly by flying around it in 1968—the moon is a sphere, and it follows that it has a back side; I don't have to see that back side to know that this is true. The same applies to speciation; give reproductive isolation long enough, and the consequence, based on the properties of variation and selection, will be speciation. The request that we prove speciation by being there at the moment "it happens" is not a rational request; in fact, it probably is designed to be impossible to demonstrate. When does a "breeze" speciate into a "stiff wind"? When does a tropical storm "become" a hurricane? When does one flower group "become" two groups? There is no hard line dividing these things, but we know they occur, and we can see it.

Second, some speciation is best viewed with a number of new technologies. This is no different than needing new technologies to understand that points of light in the sky aren't just points of light but may be entire galaxies. Considering that humanity has been seriously looking for evidence of speciation for only about 150 years, it's not surprising that we haven't cataloged thousands of cases of speciation in detail in the way evolution deniers want it to be documented. In fact, new methods, some argue (methods we'll see below), are better-suited to studying evolution than what biology has been using for decades. In a 2004 paper, genome biologist Thomas D. Kocher pointed out that "[t]raditional vertebrate laboratory models are ill suited for studies of how organisms adapt to their environment, and therefore several new [methods] are being developed to promote the fusion of ecology, evolutionary biology and functional genomics. . . . Genes will be the common currency that allows theorists and empiricists to discuss the importance of different evolutionary forces during speciation."[39]

Another method of observing speciation is often to argue that it is the only reasonable explanation for our observations. This particularly applies to the vast quantities of evidence for speciation in the form of the fossil record, the remnants of multitudes of organisms preserved through vast stretches of time. And again, we'd be foolish to argue that we must be present at the moment of an event to know that it happened (we'll see that in an example below). The city of Pompeii, Italy, has been exposed by excavating through a blanket of volcanic ash, revealing the remains of its fleeing residents outstretched in streets and alleys. There is nobody alive today who was alive in 79 CE, but it would be foolish to say that because nobody saw it happen, we can't be certain that Pompeii was covered with ash in that year, quickly and catastrophically. It would be equally foolish to say that evolution couldn't be true because biologists weren't around to observe speciation directly in the past.

Finally, it is no longer up to biologists to prove evolution to those who do not support evolution. The data are in from a dozen sciences and from studies done over 150 years; the theory works, it explains a tremendous amount, and we may accept it as fact, as something that occurs. Biological science should no more take direction from religious deniers of evolution than aircraft designers or plumbers should take such direction; life science is not the domain of religion. In fact, it is now up to the religious deniers of evolution to overturn the masses of evidence that repeatedly support evolution.

The usual critiques, then, of evolution based on pointing out the occasional difficulty we have in observing it are foolish. I've just dismantled them, and that's as much time as I will spend on them.[40]

Actually I will mention one more thing. Sometimes debating the fact of evolution degenerates into *solipsism*, when one party suggests that science is just another faith, after all, and that science can no more know anything than any other system of knowledge. First, science is not a faith; it demands evidence for claims, whereas in direct opposition, faith means to believe *without* evidence. Second, I have yet to find a single debater willing to test the idea that humanity has simply "dreamed up" the universe by, say, leaping from a ten-story window. After all, if that were true, if all we "know" is socially constructed illusion, and there is no reality external to humanity, then that leap would be of no consequence. But that challenge is never taken up. Lucretius, again, saw through this silliness a very long time ago: "If anyone thinks nothing is ever known, he does not know whether even this can be known, since he admits that he knows nothing. Against such an adversary, therefore, who deliberately stands on his head, I will not trouble to argue my case."[41]

I agree with Lucretius. Life is too short to spend on this kind of critique. I will let Stephen Gould say it best. "In science, 'fact' can only mean 'confirmed to such a degree that it would be perverse to withhold provisional assent.' I suppose that apples might start to rise tomorrow, but the possibility does not merit equal time in physics classrooms."[42]

As apples are not rising from the ground, life-forms are not fixed types; they change, and to the process of that change, we have given the word *evolution*.

Let's get back into the world of rational thought, where we know humanity has not just invented a universe around itself, but has in fact discovered some things about the universe it inhabits.

OBSERVING SPECIATION: TENNESSEE CAVE SALAMANDERS

A number of cases of speciation have been observed in the divergence (reproductive isolation) of open-air species and their relatives living inside caves. In the past, such speciations would have to be inferred from the anatomy and breeding abilities of these forms, but today we also have the ability to compare the genes of such life-forms to identify speciation.

One fascinating example comes from the caves of Tennessee. In and outside these caves there live populations of a salamander (all of the genus *Gyrinophilus*); the open-air salamanders are called spring salamanders, and the cave dwellers are called cave salamanders. Recently a Tennessean research team collected DNA from the tails of 109 of the open-air salamanders and the cave salamanders from forty-two locations. A comparison of the sampled genes showed that even though there was enough DNA similarity to show that the cave and open-air salamanders were very closely related, the cave dwellers have specific and cave-unique genetically controlled characteristics absent in the open-air populations. These include atrophied eyes, a characteristic common in cave dwellers that originated as open-air species, where light made vision a useful sense. The reasonable inference is that the cave dwellers, based on their overall physical appearance—but lack of characteristics useful on the surface—and their genes, are derived from the open-air populations; a new form of life—incipient speciation.[43]

We've seen this kind of gene comparison before, and we can ask again: if life-forms are exact types, independent, built by supernatural command to be a type that does not change (because it has a supernaturally defined function), why are they so similar to their closest relatives? It would seem that species in the open air and in caves would have quite different DNA, dedicated (and indeed perfected) to their uniquely different lives. But that is not what we see in nature. What we see are close relationships between open-air and cave-dwelling species, with the cave-dwelling species showing characteristics adapted to their own environments. Why, if these creatures were separately "punched out" by some non-evolutionary machine, would the cave dwellers begin to grow an eye that never fully develops? Only the evolutionary explanation makes sense; the cave dwellers derive from the surface dwellers, and eyes, which are living tissue that require energy to maintain, were selected against in the dark cave environment, leading to different kinds of salamanders over time. That is evolution.

OBSERVING SPECIATION: MOSQUITOES OF THE LONDON UNDERGROUND

A fascinating example of reproductive isolation leading to speciation in somewhat artificial conditions comes from the London Underground, or Tube, of all places. There, biologists Katharine Byrne and Richard A. Nichols discovered that above-ground northern house mosquitoes (*Culex pipiens*) had a subterranean counterpart in a subspecies, *molestus* (*Culex pipiens molestus*)—named for the fact that it was a particularly annoying pest to Londoners who sheltered in the Underground during World War II. The molestus form and the above-ground form differed in several ways, including the facts that molestus breed only underground, where pipiens breed only in the open air; molestus feast on mammal blood, whereas pipiens prefer bird blood; and molestus is active year-round because of the warmer conditions in the Tube, whereas in winter pipiens enters a form of hibernation called *diapause*. Byrne and Nichols collected specimens of the Tube dwellers and the surface dwellers and tried to mate them; you can imagine, by now, that the divergence between the forms since the Tube (opened in the 1860s) was colonized by what would become molestus resulted in these matings being unsuccessful. Furthermore, genetic tests showed that the two populations were very closely related. Could the subterranean populations simply be immigrants from somewhere else, rather than derived from the surface population? Recent genetic studies suggest that they might be related to other mosquitoes from warmer regions, but what isn't in question is that the subterranean forms have adapted and become reproductively isolated in the Underground. Here, then, is a case of reproductive isolation in what the biologists estimated to be a few hundred generations, as well as speciation.[44]

OBSERVING SPECIATION: CLAM WORMS AND FRUIT FLIES

The mechanisms of speciation have been observed in labs across the globe, and it's arguable that a few cases of speciation have in fact been engineered in the lab. Saying that the concept of speciation is false because of this (a) ignores the fact that speciation has been observed outside the lab and (b) ignores the fact that many processes are easier to observe and document in a lab than in the field, for all kinds of logistical and time issues.

A relatively early observation of speciation in laboratory populations is seen in the case of the clam worm *Nereis acuminata*, a bristly bottom dweller. In the early 1960s, a handful of these worms were collected from coastal California and shipped to the Woods Hole Oceanographic Institution in Massachusetts, where the population grew to thousands of individuals. Nearly thirty years later, more worms were collected from the Californian waters, and researchers then attempted to breed them with members from the Woods Hole population, which, as I mentioned, had been totally isolated from the Pacific populations for nearly three decades. The results were striking; in over one hundred tests, when matings between members of the California and Woods Hole populations were attempted, none was successful. The populations had diverged; speciation observed in a laboratory.[45]

In another case, Greek biologist George Kilias collected a population of six hundred members of the fruit fly (*Drosophila melanogaster*) from the wild and then segregated them in laboratory boxes for several years. Half the flies were housed in cold, dark conditions, and half in warmer, lighter conditions, and the flies feasted on cornmeal provided by their human caretakers. When Kilias opened the boxes and "introduced" males and females reared in different (cold/dry, warm/wet) conditions, the cold/dry flies were loath to mate with the warm/wet set, and vice versa. Remember premating reproductive isolation, in which members of populations of the same species simply won't mate with one another? Here it was, developed over about eighty fly generations in the laboratory. Is there any reason such isolation and differentiation could not occur in nature?[46]

OBSERVING SPECIATION: STICKLEBACKS WORLDWIDE

New molecular methods have completely revolutionized testing of everything ever predicted by evolutionary theory about speciation. The reports are pouring in, and they verify that Darwinian evolution is essentially correct.

One spectacularly well-reported example is known from many

studies of the stickleback, a small, feisty fish widely distributed in the Northern Hemisphere. Sticklebacks are known in both oceanic and freshwater forms. Geological studies indicate that many freshwater populations were isolated in lakes and other water bodies as global sea levels dropped significantly at the end of the last ice age, something on the order of ten thousand years ago. When this happened, freshwater and oceanic forms became significantly reproductively isolated. What happened next? You can imagine; the freshwater forms began, through variation and selection, to change in ways that made them more fit for their new freshwater habitats. Freshwater sticklebacks lost the armor plating of their oceanic ancestors, as well as their "pelvic spines," the oceanic stickleback's sharp, downward-pointing defensive features. Exactly why losing these made them more fit is under investigation, but it may simply be that the armor plates and spines are both living tissue, and if conditions in freshwater don't penalize a stickleback born without them, then the lowered energy cost of not having to "feed them" might act as selection "against" them. In other words, freshwater sticklebacks, for whatever reasons, could afford to be born without the defensive features of their oceanic ancestors. Exactly why is a point of research right now.

What is most striking is that this happened *independently* in many populations; for example, in Iceland and British Columbia. In both places (and in Japan and others), exactly the same loss of pelvic spines and armor plating happened when the fish were isolated in freshwater habitats. Not only that, but it happened because of the same genetic variation; in the case of the pelvic spines, a gene called *Pitx1* has been found to differ in the oceanic and freshwater forms. *Pitx1*, known from many species, is important to the development of the hindquarters in many animals—the legs in mammals, for example, and the pelvic spines in sticklebacks. Apparently this same gene has varied in each of these independent populations in the last ten thousand years in the same way, resulting in more armorless and spineless sticklebacks in freshwater. All of this, again, is known to have occurred in the last ten thousand years because the lakes and other freshwater bodies occupied by the freshwater sticklebacks didn't even exist before this time; ice covered the land.[47]

Not only can we observe change in the body shape through time, then, and the gene responsible for it, but we can also observe that strong reproductive isolation has appeared among sticklebacks even *within* the freshwater lakes. In one study of British Columbia lake stickleback, those that prefer to live and feed at shallower and deeper waters of the lake strongly preferred to mate with their own kind, that is, dwellers of one's own (from the stickleback's point of view) depth. That reproductive isolation hasn't yet led to naming the deeper and shallower sticklebacks new, different species—remember, science is slow to move. But by all evidence, they do indeed seem to be different species, diverged in the last ten thousand years as replication, variation, and selection shaped each for the selective environment of a different depth. That is evolution.

OBSERVING SPECIATION: BAHAMIAN GUPPIES AND AFRICAN LAKEFISH

The best studies use several different lines of evidence to see whether each supports or refutes the hypothesis of speciation . . . and again and again the results are the same; speciation happens. One such study used behavioral, body size, and genetic studies to investigate populations of mosquitofishes of the Bahamas; the populations, each inhabiting separate "blue holes"—circular bodies of water on the islands—were found to have diverged in the last few thousand years, often shaped by the kind of predation they faced in their specific blue hole.[48]

In the past decade, numerous cases of speciation have been revealed by a multitude of studies of the cichlid fishes of East Africa. Geological evidence shows that in many cases—as in Lake Victoria—cichlids colonized lakes in recent geological history. In Lake Victoria, specifically, an "adaptive radiation" apparently took place in the last twenty thousand years, in which the original founding populations of fishes "radiated" (fanned out) into the many habitats that were new to the founders and adapted to each, leading to over five hundred cichlid species with unique DNA, feeding habits, behavior, and so on.[49] Once again, the genetic evidence indicates that these are recently related life-forms that diverged from a common ancestor; and again we can ask whether it is more likely that these many closely related species "popped up" out of nowhere—a process never seen before in nature—or are simply variations on a common theme, as reflected in their genes, shaped over time by replication, variation, and selection in slightly different subhabitats of Lake Victoria.

Genetic and anatomical reconstructions of the history of the Lake Malawi cichlids, where speciation took longer, reveal at least two natural stages to speciation: first, cichlids adapted to either rocky or sandy habitats; following this, due to differences in food in these habitats, selection "shaped" differences in "feeding apparatus" (mouth, teeth, and jaws), which diverged into forms such as "algae-scraping" and "suction feeding."

And a recent genetic study shows that cichlid jaws—one of the characteristics that have changed the most during speciation as different populations adapted to different foods and different kinds of eating—are controlled by a handful of genes; again, we can now say exactly *how* these changes occurred.[50] We don't have to generally wave our hand at the water and say, well, somehow these fish forms changed through time; today we can point specifically to individual genes—genes that make proteins and genes that switch other genes on and off, regulating development—and say, *there*, that is what happened, that base pair or codon (or what have you) was altered by whatever source of variation, and that variation was, for whatever reason of the selective environment, selected *for*, leading to different kinds of life than there were in the past. That is evolution.

OBSERVING SPECIATION: VOLCANIC CRATERS OF CENTRAL AMERICA

Let's see just one more of the many recent studies observing speciation in fishes (and other life-forms) that have recently populated new habitats and, in those habitats, speciated into different forms. This case comes from cichlid fish (again) in a Nicaraguan volcanic crater. The crater, dated by geological methods, was formed in a volcanic explosion about 23,000 years ago and since filled with rainwater (to a depth of more than six hundred feet) and has been colonized by two kinds of cichlids that are completely reproductively isolated: the smaller arrow (*Amphilophus zaliosus*), which has pointy teeth, and the fatter midas (*Amphilophus citrinellus*), which has flatter teeth. Analyses of stomach contents of these fish showed that the arrows dine more on insects, while the midas prefer algae and other plant material. While these two populations differ in their body form and diet, and they do not interbreed, genetically they're very similar, and—very tellingly—when compared to arrow and midas cichlids in six *other* Nicaraguan lakes, they were found to be very different from those other populations. The conclusion is that the lake was in the last twenty thousand years or so colonized by a cichlid that later diverged into two forms with different body shapes, diets, DNA, and breeding habits: speciation.[51]

Why isn't the interpretation that the lake was initially colonized by *two* forms? Because genetically they're so close as to be almost indistinguishable, *and the genetic distance between them is much smaller than the genetic distance between other species of cichlids.* Can we use the data in an aircraft's black box to identify what happened in the past, even without being there to witness it? Yes. Can we use the genetic data in these fishes to identify what happened in the past, even without being there to witness it? Yes. This is speciation, written in the genes.

OBSERVING SPECIATION: MONKEYFLOWERS OF THE WEST

There are over a hundred species of monkeyflower, a colorful plant that occurs throughout the western United States and Canada. Reports of recent speciation in some Oregon populations came in as early as the 1970s, but more recently a lot more investigation—and a lot of it using modern molecular tools—has demonstrated speciation in monkeyflowers. One study compared the monkeyflowers from higher and lower elevations in California. Although the two plants are genetically very similar, the monkeyflowers from the higher elevations are pink, and they're pollinated by bees, while the lower-elevation monkeyflowers are red, and almost all their pollination is courtesy of hummingbirds, not bees. This difference in pollinators is a strong element of reproductive isolation; higher-elevation bees simply don't carry pollen down to the lower monkeyflowers, and lower-elevation hummingbirds only rarely go up and pollinate the higher-elevation monkeyflowers.

These monkeyflowers aren't completely separated; a few hybridizations—crosses of the higher- and lower-elevation forms—do happen in the elevation range between the higher and lower monkeyflowers. But the hybrids are not very healthy and produce fewer seeds than usual. The genetic similarity between these monkeyflowers shows their essential relationship, but strong reproductive isolation has reduced gene flow between the two populations. Currently these forms are considered different species (*Mimulus lewisii* at higher altitudes, and *Mimulus cardinalis* at lower altitudes), and based on the same genetic similarity we see in such cases as the Nicaraguan cichlids (just discussed), that difference is reasonably interpreted as a speciation resulting from the fact that different pollinators service different elevations. That is evolution. If it isn't, what should we call it?[52]

OBSERVING ANCIENT SPECIATION

Before the modern molecular evidence was even available, speciation was observable in the study of fossils, remnants of ancient life-forms, all around the globe. To understand how they're interpreted, we need to know a little about fossilization as well as the dating of these fascinating glimpses into the very ancient earth.

When a plant or animal dies, it may or may not be entirely consumed and broken down by various forces. Sometimes, plants and animals are preserved by various processes, the best known being *fossilization*, the replacement of the organic tissues with minerals. As the body (a leaf, a bone, or other tissue) breaks down, during fossilization small mineral particles replace the organic tissues. This can happen, for example, when a plant or animal dies near a lakeshore and, rather than being devoured or simply eroding away, it slips into deep mud or is covered up by silt and sand. Those sediments protect the bones or leaves, and over time the minerals in the sand and silt actually migrate into the cavity left by the decaying organic tissues. Fossils can be very high-quality casts; they even preserve scratches and other microscopic wear on fossil teeth. Fossils, then, aren't strictly remains of the life-form but high-quality casts of ancient life-forms.

There are plenty of ways to determine the age of a fossil, and this book isn't the place to review them.[53] The essential fact, though, is that dating methods are well tested and reliable. One way to test our methods is to take a sample of a known age and send it to a lab, without telling the lab the actual age; and lo and behold the dates come back correct. With radiocarbon dating, for example, I can divide my sample in three and send one piece to a lab in Australia, one to a lab in England, and one to a lab in California, and with no collusion on their parts—they don't even know I'm doing the test, only that I'm paying them to "run" the date—the dates come back the same. Could the labs all be making the same mistake, though, all misdating the material? No. Pieces of wood from ancient Egypt, dated by historical documents to, say, three thousand years ago, can be sent to independent labs, and the date comes back correct and the same from each lab. The methods are secure.

Occasionally we do need to revise our dates (just like, occasionally, an airplane does crash even though we know how to make them fly), but it's rare that major errors are made, and when they are, they're quickly caught. This is because the geologists working with fossil hunters use not just any single method to date a fossil discovery but many; the use of multiple, independent lines of evidence.

One of the oldest dating methods simply identifies the *relative* age of a fossil. If the geological processes of the past were essentially the same as today, and a fossil site hasn't been disturbed, then the fossils deeper in the ground will be older than the ones nearer the surface. This is because over time, due to gravity and erosion, sediments stack up in layers. If we know a fossil comes from deeper in the ground than another, then, we can start to fit together a sequence, and since changes in populations over time is what happens in evolution, time sequences are critical. This is a crude but effective method used throughout the nineteenth century; by the 1950s, however, new technologies allowed the *absolute* dating of the layers; these dates indicated, in years, how much time had passed since some event, like the solidification of lava into rock. This gave science a much finer-grained understanding of the passage of time and the relationships of certain fossils to one another.

Since species can—at large—be determined by their outward form (their shape), and since we can date the fossils of ancient species, it follows that we can trace the history of a given species over time. One proviso; since the commonly (though not universally) accepted species concept—Mayr's statement that species are populations that breed among themselves—depends on whether or not two of any population can actually breed, for fossil species, of course, that cannot be tested; we can't try to breed two members from the distant past because their organic tissues have decayed. But we don't throw the baby out with the bathwater. We can use the same inference that detectives use to solve murders: with multiple, independent lines of evidence we can often be quite sure about what happened in the past.

In the case of observing speciation in the fossil record, we can be sure that speciation explains the fossils in the same way that speciation explains our observations about the living world. That is, in the living world, we see replication, variation, and selection, and we do not see species simply popping up out of the blue. It would be perverse to suggest that the same replication, variation, and selection did not occur in the past. Of course it did; it was driven by the same DNA that drives living things today; ancient DNA studies, which we'll revisit later in this book, have discovered DNA that is millions of years old.

How do we know that speciation happened in the fossil record? Basically, when we see a change in the fossil species of a magnitude that is comparable to the differences between living species, it is clear that speciation has occurred. See that change over time, and it's a good bet that you're looking at some kind of reproductive isolation followed by speciation.

Actually, we don't even need to be able to point to a certain pair of fossils and say, "This is species X and this is species Y" to understand that evolution has taken place. Why not? Because the actual lines around a species at any given moment are less important than the fact that *species change over time*, eventually leading to new kinds of life. To paraphrase Dobzhansky, quoted earlier in this chapter: *a species is a phase, not an end result.*

OBSERVING SPECIATION: ANCIENT PLANKTON

One of the best examples of observable speciation in the record of ancient (fossil) species comes from core samples recovered from layers of seafloor sediment in the equatorial Pacific. Geological dating methods indicate that these samples represent a span of at least 2 million years, from 3.4 million years ago to 1.6 million years ago. A study by biologist Ulf Sorhannus revealed that the sediment cores contained millions of tiny, cone-shaped plankton of the genus *Rhizosolenia* (millions of living *Rhizosolenia* are found in a single pint of seawater.) Sorhannus and his colleagues studied over five thousand of these diatoms under the microscope, discovering that in the oldest levels (about 3.5 million years ago) the plankton's "hyaline area" (an extension of the tip of the cone related to feeding) was around four microns (much smaller than the thickness of a human hair) in size. But among the middle core samples that represented about three million years ago, many of the diatoms had a somewhat smaller hyaline area, although plenty were still around four microns in size. And in all the core samples representing any time after three million years ago, there are clearly two very different populations; one with about the same hyaline area as the diatoms 3.5 million years ago, but another kind—completely new—with a hyaline area roughly four times *smaller* than that of the oldest *Rhizosolenia*.[54] The anatomical difference between the older and more recent of these diatoms—the difference in hyaline area—is comparable to the differences we see in modern closely related (but different) species, and it is reasonable to say that, failing any other explanation, the ancestral population (of around 3.5 million years ago) for some reason diverged around 3 million years ago, leading to two new forms: speciation.

I just said "failing any other explanation." *Is* there another explanation? Is there any mechanism known to science that "punches out" new life-forms out of the blue, over time laying them down with the remains of other species? No. In the same way it's reasonable to reconstruct a murder from the evidence at the scene—even though we weren't there to witness the event—it's reasonable to say that somewhere around three million years ago, speciation occurred in this life-form.

OBSERVING SPECIATION: FOSSIL ANCESTORS

Perhaps the most compelling evidence of speciation is seen in the fossil record of the *hominins*, the large bipedal primates. There is only one of these species alive today (you and me, *Homo sapiens sapiens*), but there have been others in the past. Hundreds of fossils of early hominin bones dating from about six million years ago to about one million years ago have been excavated in Africa. The world's best geologists have spent decades dating sediments in which the fossils are found, and while minor squabbles about details erupt now and then, everyone agrees on the basic chronology. This has resulted in an extremely fine-grained chronology (time sequence) based not on just one method or one study but on hundreds of studies using about half a dozen independent methods.

Early in the sequence, we see fossils that look somewhat like the chimpanzee and somewhat like humans; later in the sequence, we see something looking much more like humans, and not just in the shape of the skull and jaw but in the entire *locomotor* skeleton, the skeletal elements related to locomotion, or getting around. By three million years ago, it's clear that some of these primates are entirely bipedal, walking on two feet. And again, independent methods confirm what we see in the fossils; the Laetoli footprints, a trackway of footprints found by paleoanthropologist Mary Leakey (1913–1996) in the early 1970s, are dated to over three million years ago, and only bipedal primates could have made them. And the genetic clock shows that humans and chimpanzees diverged somewhere around six to eight million years ago.

By two million years ago, the braincase of some of the hominins are dramatically larger than the earlier forms, and by a million years ago, one of the hominins—one of these large, upright apes—simply vanishes: extinction, no more of them. But while that hominin became extinct, another hominin, similar to modern humans in its large braincase and small teeth, continued. We see its fossils continue in the record, up through the increasingly more recent geological layers. At some point—and anatomists debate exactly when we should "make the call" that shows the "arrival" or "appearance" of the first member of the genus *Homo*—that line of hominins has a brain distinctly larger than any before it and teeth far smaller than any before it, and archaeological data show that it was using stone tools to survive. We see in this fossil record, then, of hundreds of bones preserved over millions of years, a gradation of a form that looked, early on, like the chimpanzees who are our closest genetic relatives, toward the modern human form. Whenever we want to draw the line to say that *Homo* first appears, the fossil record clearly shows a continuation of earlier trends, with a brain growing larger, teeth becoming smaller, and an ever-increasing reliance on tools to survive. This is the evolution of our genus, *Homo*. It leads right up to forms that, by one hundred thousand years ago, cannot be distinguished from modern human populations; that is called *anatomical modernity*, and it effectively means the "appearance" of modern humans—but remember, they did not appear from nowhere; hundreds of fossils show that they—like other species with good fossil records—have an evolutionary past.

Again, we can ask: do we see anything in the natural world that would punch out new, unrelated life-forms like these different forms of *Homo* and drop them in the fossil record over time? No; we see—most often—the gradual transformation, or evolution, of one life-form into another.[55] And if there *were* such a mechanism, why would it drop these fossils in a sequence that diverges more and more from the chimpanzee-like form early on and more and more toward modern humans? There is no reason for that; this hypothetical force could just as easily punch out forms that never changed, or that changed toward, let's say, the form of some other species. But the fossils don't; they change in a way that we today, with hindsight, can clearly see is from one species to another—these hominins being as different from one another in the fossil record as living species are from one another, today, all around us—toward the modern form of humanity.[56]

We have seen many cases of speciation here, and these are just a few. We have observed speciation and its processes in genetic, behavioral, and anatomical studies of living populations, in laboratory populations, and in the fossil record. It is now the evolution denier's burden to show these are *not* cases of speciation, that some other natural process can account for the changes, over time, in these many life-forms past and present, that lead to new species.

SPECIATION AND EVOLUTION

We have seen the replication of life-forms and their membership in breeding groups—naturally segregated populations that we can call species. These are obvious and observable even in the record of ancient life-forms. But these facts alone don't explain the millions of life-forms around the globe today, like the diversity of life-forms on the cover of this book. Why is there more than one kind of fish, for example, or plant, or land animal, or bird? When we consider that populations are subjected to changing selective pressures, and that as they spread across the globe new populations encounter new selective pressures, the answers are clear; subpopulations of original populations become reproductively isolated, for geographical, anatomical, or behavioral reasons, and their own selective pressures change them over time into new species: the fact of speciation.

The only alternative is *independent creation*, the idea that all these millions of species simply "popped up," but there is no scientific evidence for that at all in the fossil record, living memory, written history, or our daily experience of the world we live in. Rather, we see a long and rich fossil record that records the rise and fall of many millions of species over the past four-plus billion years of Earth life. And new methods, such as DNA analysis—the same methods that we use to identify paternity, convict people and send them to death, and exonerate others—show that, yes, the species we draw lines around (however sketchily, because sometimes species "leak" a few genes into other groups) are real, and yes, species have histories that support what the fossil evidence tells us.

The fact that the fossil and genetic evidence corroborate one another—even though they were developed centuries apart and for different reasons—is compelling support for a Darwinian explanation of the diversity of life today and in the past. To dismantle evolutionary theory today, deniers of evolution are going to have to dismantle molecular biology as well as the study of ancient life in the fossil record. That is a tall order.

CHAPTER 6

THE FACT OF EVOLUTION

> **So many atoms, clashing together in so many ways as they are swept along through infinite time by their own weight, have come together in every possible way and realized everything that could be formed by their combinations. No wonder, then, if they have actually fallen into those groupings and movements by which the present world through all its changes is kept in being.**
>
> —*Lucretius: On the Nature of the Universe*[1]

We've seen, now, how replication, variation, and selection occur. Each is observable in nature. The consequence of these processes—change in the characteristics of life-forms over time that lead to new forms of life and the diversity of life across the globe—is what we call evolution. This is all very simple and irrefutable; evidence for each is observable in nature every day with the tools we've seen again and again in this book.

If someone says they do not believe in evolution, one may very reasonably ask whether it is replication they don't believe is happening, or variation, or selection. If those are given—and to any rational thought, they have to be, because they are observable facts—then evolution is given because it is simply the *consequence* of these facts.

Fig. 6-1. Evolution as the Consequence of Replication, Variation, and Selection

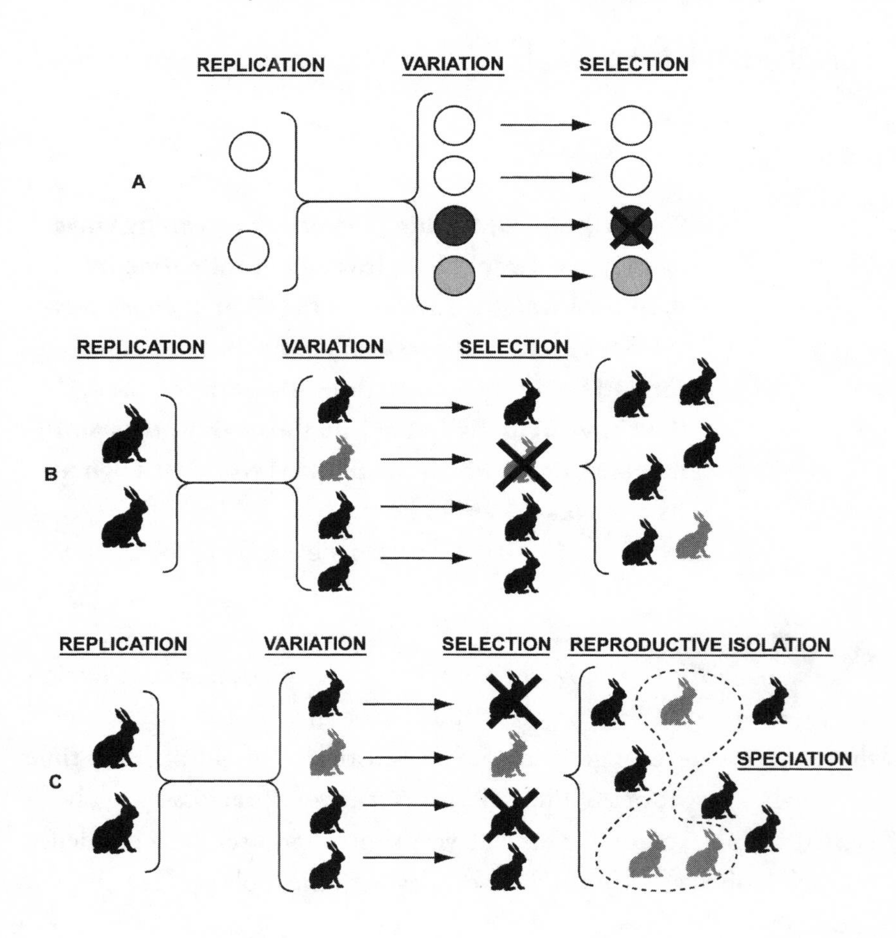

Figure 6-1 schematically shows replication, variation, and selection in action. In panel (A) we see "Replication" as the mating of two life-forms. Under "Variation," we see that while the offspring are largely similar to their parents and one another, two have a variation; the bottom two are shaded. And under "Selection," we see selection; for whatever reason, the darkest-shaded offspring is not well suited to the environment (it is less fit than its companions), and it does not survive long enough to pass on the genes for its dark coloration. In contrast, while the upper two offspring survive, another variation, the medium-gray shaded circle at the bottom, also survives and passes on the genes for that variation. Over time, if that variation is beneficial, it should become common in the population. In the central panel (B), we see a case where two parents mate, having four offspring that are close copies, but with one lighter-colored variation. For whatever reason, that lighter color is deleterious—it reduces fitness, somehow making that individual a little less well-suited to its environment. For this reason, under "Selection," the individual carrying that variation is selected against (X), and the genes for that variation do not continue in the gene pool or will be very rare (in the event it lives long enough to have a few offspring). Finally, in panel (C) we see a case where a lighter variation resulting in the offspring of a pair of replicators is selected *for*, for whatever reason that lighter variation conferring some kind of advantage. In contrast, the darker-colored offspring, who once were in the majority and were the most fit, might be actually selected against if, for example, the environment changes and being born with characteristics that worked in *their parent's time* are no longer beneficial. With the selective pressure now changed, the genes for lighter coloration become more common in the population, because they confer a greater fitness. Over time, if reproductive isolation arises—for whatever reason the darker and lighter forms no longer interbreed (dotted line)—the populations might diverge so much that speciation occurs; there are now two kinds, a light and a dark species, whereas before there was only one.

Let's wrap this up.

Replication

Life-forms are born of parents, resembling them closely because they are built from their parents' information-rich DNA.

Variation

Life-forms vary (for several reasons), and those variations result in individuals having different chances of having their own offspring.

Selection

Variations that are beneficial tend to be passed on (they are *selected for*), and variations that are not beneficial tend not to be passed on (they are *selected against*).

Changes in Selective Pressures

Because environments change over time, exactly what variations (characteristics) are beneficial to survival will change over time, changing the characteristics of a life-form (species) over time.

Reproductive Isolation

If individuals of populations diverge in their habits or environments, they might cease to interbreed freely with others of even their own species.

Speciation

If members of a population persist in not breeding with others of that same population, the two populations may become so adapted to their own environments that they can no longer interbreed, resulting in two kinds of life when previously there was just one: speciation.

And that is that. As we've seen, there is almost no *there* there with regard to "evolution." As humans with hindsight, we can say that these things occur over time, draw a line around the process, point to it, and call it "evolution." But I think you can see that the "itness" of evolution, the "thingness" of it, is an illusion. "It" is simply the logical consequence of replication, variation, and selection. It is like a golf ball let go at the top of a slope. Because the ball is not attached to anything, and because the surface on which it is let go is a slope, and because the ball is of less mass than the earth and it is not moving fast enough to escape the gravitational draw of the earth, the ball will move. And the consequence of these circumstances is that the ball will roll down the slope. Evolution is the ball rolling down the slope; it is a consequence, not a plan or even a "thing." It just happens. Because of the replicating, variable nature of life-forms, and the dynamic, changing nature of the universe, evolution *has* to happen.

With this perspective, life-forms around us take on an entirely new dimension; looking at a leaf, a squid, a fly, a tree, a bee—thinking of a life-form as a sort of historical document opens a dozen doors of new fascination:

- What is the history of that life-form?
- Are its characteristics "perfectly adapted" for the conditions we see today, or could they be vestiges from another time?

- What is that life-form's selective environment? What are the selective agents that affect its fitness?
- What species does this life-form co-evolve with? Does it have parasites? Is it a symbiont? Don't forget the microbial world!
- How long can that life-form survive, considering changes in the world?

CHAPTER 7

EVOLUTION IN ACTION

> **[The crocodile] lives in the water, its mouth is all full of leeches. All birds and beasts flee from it, except the sandpiper, with which it is at peace because this bird does the crocodile a service; for whenever the crocodile comes ashore out of the water and then opens its mouth (and it does this mostly to catch the west wind), the sandpiper goes into its mouth and eats the leeches; the crocodile is pleased by this service and does the sandpiper no harm.**
>
> —Herodotus, ca. 400 BCE[1]

We've now seen the three facts of the natural world that inevitably result in what we call evolution. The best way to appreciate this, short of going out into the wilderness to observe it, is to learn about how life-forms live in the wild, keeping an eye out for how replication, variation, and selection play out in natural populations. In this chapter, I've selected a handful of observations about a number of life-forms that should make this clear. These are sketches of natural life that allow you to appreciate the complexities lying behind the essentially simple processes of replication, variation, and selection, and how they result in evolution. Although replication, variation, and selection might not be explicitly discussed, you can imagine how they are playing out.

These accounts also give you a good impression of how life scientists learn about natural-living evolutionary processes in the first place, in dedicated field studies across the globe and over years, decades, and entire careers. Remember, in each case, the scientists studying these natural phenomena have to know about the species' life course, its food sources, predators, temperature tolerances, and so on; whole selective environments, as complex as they are, have to be understood. Few are completely understood. Science is a work in progress. No wonder field studies, the analysis and writing up of results, and the dissemination of these results to the scientific world as well as the general public can take a long time.

Finally, keep in mind that in every case, if Darwinian evolution by replication, variation, and selection were fundamentally incorrect, wouldn't that pop up, wouldn't it be obvious somewhere in the many thousands of such studies that have been done over 150 years and across the globe?

EVOLUTION IN ACTION: MEERKAT MORTALITY

African meerkats are gregarious, cooperative mongooses. One study over four field seasons of close observation revealed much more about the world of the meerkat than we could get in any quick glance from a Land Rover.

Meerkat breeding is seasonal, with most births taking place during the rainy season. And meerkats live fast: females are ready to mate again just a few weeks after giving birth. Meerkat young are called kittens, and, as in many species, the young are in particular peril. Many don't survive past three to five weeks of age, and about 30 percent die from predation or cold in the first thirty days.

Those that make it past this age still have to survive fluctuations in rainfall (which affect the food supply of plants and insects) and variable time of female care of the kittens; if females are themselves in poor health, they invest less time in caretaking, leaving kittens vulnerable. Why might the caretaker females be stressed? Because just as they are coming out of the difficult period of gestation—when they have to obtain a lot of food—they come into the period of lactation, when they also need a lot of food. Kittens are also occasionally killed as pawns in a power play of the social order, when higher-ranking females kill the kittens of lower-ranking females.

Meerkats have evolved a social order that includes the cooperation of helpers that protect offspring as best they can, sometimes bringing them food, or tucking them away in their dens if predators come around. It doesn't always work. In one case, an entire litter of kittens was lost when a tawny eagle (*Aquila rapax*) continually harried the den, preventing both the foraging adults and the helper—who had finally left the den after going two days without food—from returning. The eagle didn't actually kill the kittens; they succumbed to hypothermia in the den, despite their habit of huddling for warmth (something over twenty mammalian populations are known to do). Other times, predators are effectively driven off. Once, a Cape cobra (*Naja nivea*) entered a den and ate two kittens before it was finally driven away by a mob of adults. In addition to predators and cold, kittens are also vulnerable to drowning if the den is flooded. In one case, though, a helper meerkat singlehandedly moved an entire litter from a flooded den to a dry den, over 50 meters (165 feet) away, while the rest of the band were out foraging.

If kittens survive to about twelve weeks, they're normally mature and healthy enough to forage out beyond the immediate den area. Then they enter the world of foraging, mate seeking, and complex social relationships.[2]

Meerkat kittens are born into daunting selective environments in which weather, social relations, predators, and even the health of their own parents and of the group at large play a role. To survive, meerkat kittens and their caretakers all have to have characteristics—variations—that allow them to negotiate all these selective pressures.

EVOLUTION IN ACTION: WASPS AND ZOMBIE COCKROACHES

Species, we have seen, do not exist in a vacuum; they interact with other species, be they prey or predators, vegetation, parasites, or what have you. One particularly interesting case is that of the jewel wasp (*Ampulex compressa*), which has a parasitic relationship with the cockroach (*Periplaneta americana*). When a jewel wasp spots a cockroach, it zooms in and stings the cockroach in the head. Usually, then, the cockroach will quickly enter a *hypokinetic* state; *hypo* meaning "reduced," and *kinetic* referring to motion—it stops moving or slows down significantly. Its antennae move just a little, whereas they would normally move almost continuously as the cockroach senses its environment. This behavior is in great contrast to un-stung cockroaches that scuttle around in a "hyper-exploratory" way. Next, the wasp grasps the cockroach's antennae and, to quote the original scientific report, walks it "to a nest much like a dog on a leash."[3] At the nest, the wasp lays an egg in the body of the cockroach, seals up the nest, and goes about its business. When the egg hatches, the hatching wasp devours the hapless, zombified cockroach.

How can this type of behavior persist? Wouldn't the cockroach population be wiped out by wasp parasitism? It would, if every cockroach were so affected, but not all are; some have a resistance to the wasp venom. In a laboratory study, just under 15 percent of stung cockroaches were able to respond to stimulus (and, in the wild, presumably, survive the injection). Also, not every wasp is successful in injecting its target cockroach, and cockroaches are very numerous. How long has this relationship been going on? Is this relationship entirely in favor of wasps, or could the cockroach population actually benefit in some way?

EVOLUTION IN ACTION: HERMIT CRABS AND THEIR SHELLS

On the San Juan Islands off the Pacific Northwest Coast, hermit crabs scuttle at the shore. They don't grow the shells they inhabit; those are (inadvertently, of course) provided by sea snails: when the snail dies, the hermit crab moves in and then carries the shell on its back as it moves around in search of food and mates. One three-year study focusing on the relationship of the shell-providing snails to the hermit crabs tagged over four thousand snails on a 200-square-meter (2,152 square feet) study area. Observations and careful recordings indicated that about half of the snails died each year, making available a multitude of shells 20 to 40 millimeters (1.1 inches) in length. With a population of about two hundred hermit crabs in the study area, about half to one shell became available per crab per month. This gave the crabs some choice in their shells (which they inhabit until they grow too large, at which time they have to move out and find a new shell). Interestingly, the hermit crabs picked carefully among these shells, selecting thicker ones over thinner ones, probably because thicker shells, though they weighed a bit more, provided greater protection than thinner shells. The crabs also rejected broken shells.

Complications arose, however. Wave action removed shells that weren't quickly inhabited, and more shells were available in spring and summer than in the winter.

Overall, the hermit crab population was limited by all the usual selective pressures, such as predation and temperature sensitivity, but also by the availability and quality of snail shells to use as shelter.[4] What if those shell-providing snails were to move away, driven off by some selective pressure completely independent of the snails and the crabs? Is there enough variation in the hermit crab population to deal with such a change? Would some crabs, for example, based on some variation in how they process information, find some other kind of shelter? Would they migrate out of the area? Would that population simply die off?

EVOLUTION IN ACTION: HUMMINGBIRD HABITATS

In the Guadalupe Canyon of the Arizona-New Mexico border region, a four-year study revealed that hummingbirds living close to one another picked very specific places to make their nests; again, something a simple glance at this beautiful desert landscape with its flitting hummingbirds would probably miss.

To remain healthy enough to fly to flowers and hover while extracting their calorie-packed nectar, these hummingbirds had to find high-quality energy sources to fuel their flight and hovering, and the amount of energy depended on their body weight. Since the hummingbirds had different average body weights—violet-crowned species weigh almost 6 grams (0.2 ounces), and most others weigh about 3 grams (0.1 ounce)—they had different caloric requirements. It's no surprise that the hummingbirds chose to nest near the best stands of nectar-providing flowers. But not all the hummingbirds could nest in exactly the same spot, so a variety of very subtle differences arose in their habitat preferences.

Black-chinned hummingbirds nested about five meters (sixteen feet) up in Arizona sycamores (*Platanus wrightii*) that grew above bare creek bottoms. Violet-crowned hummingbirds (*Amazilia violiceps*) preferred drier, more open areas and nested about seven meters (twenty-three feet) up, also in Arizona sycamores. Broad-billed hummingbirds (*Cynanthus latirostris*), however, preferred to nest low: just a meter or so (about three feet) above the ground, sometimes near rocks; they also preferred to nest on the northern slope of the canyon. Finally, Costa's hummingbirds (*Calypte costae*) preferred to nest in the dry arroyo tributaries adjacent to the main canyon or on the south side of the canyon, about one to two meters (three to six feet) above the ground; they preferred small trees such as the netleaf hackberry (*Celtis reticulata*).

In the confines of just this canyon alone, then, four kinds of hummingbirds competed for nectar. Even though they occasionally battle for access to nectar—with the violet-crowned species being most dominant and aggressive—these species generally leave one another alone. This isn't because of some kind of group decision or committee, but because different kinds of hummingbirds have, over time, "settled" into these different micro-habitats.[5]

Did these species all begin as one and diverge into their current micro-habitats?

EVOLUTION IN ACTION: GUPPY ADAPTATION

One experiment in natural conditions showed that natural selection could change the characteristics of a species very quickly. California biologists David A. Reznick, Heather Bryga, and John A. Endler (author of the thorough book *Natural Selection in the Wild*) observed that in one Trinidadian river, guppies (*Poecilia reticulata*) of large size were preyed on by the pike cichlid (*Crenichla alta*). Knowing that other fishes (namely the killifish *Rivulus hartii*) subsisted on significantly smaller prey, the team transferred two hundred guppies from their home environment (where larger guppies were threatened by predation) to another river where killifish preferred smaller prey. Several studies over the next eleven years (about fifty guppy generations) measured the guppies and the speed at which they matured and grew to their full adult size.[6]

Within about seven years, the transferred guppies were giving birth to significantly larger offspring, and those offspring were maturing faster. The result: guppy offspring spent less time in the smaller, predator-vulnerable size. And, laboratory tests showed that both maturation time and overall body size were genetically controlled, not just environmental effects. The inference—as secure an inference as any detective would make while assembling a murder case—was that killifish predation shaped the life history of the guppies in about fifty generations.[7]

One way to say this is that there was selection for faster maturation and larger body size in guppies "by" the killifish. Another way to say it is that there was selection *against* smaller body size and longer maturation period, both of which would expose the guppies to predation. On a significant level, it really doesn't matter how we say it; this is evolution by replication, variation, and selection. Presumably some of the transplanted guppies *did* continue to produce their usual small, slow-maturing offspring after introduction, but after a while, those were wiped out, and the survivors were the fitter: those that matured faster and grew to a larger body size and then passed on their genes for larger body size and faster maturation than their peers. That's evolution, and not just by some vague "adaptation" to the environment but by very concrete, clear steps we can observe and understand. Extend this process over time, and you will get significant changes in the species and perhaps even speciation, a reproductive difference between the guppies preyed on by larger- or smaller- preferring predators.[8]

EVOLUTION IN ACTION: POLLINATOR CHOICES

It's often assumed that pollinators—animals that inadvertently move sperm-bearing pollen from male plants to egg-bearing female plants—rather simply and instinctively react to flower color in their search for nectar. But a study based at the Rocky Mountain Biological Laboratory showed that things are more complicated.

In a variety of tests that artificially changed the colors of flowers (by painting new colors on them), it was revealed that when Rufous hummingbirds (*Selasphorus rufus*) learned that pale flowers could provide more nectar than red flowers, they switched their preference to feeding on pale flowers. And when color was eliminated as a variable for hummingbird feeding decision by painting all flowers the same color, in about fifty feeding sessions the hummingbirds made a new choice, preferring to feed from flowers with a wider *corolla* (the "feeding tube" that admits the hummingbird beak), which made feeding easier.

Hummingbirds, then, are not simply slaves to instinct; they can learn about changes in their environment and adjust their behavior, recognizing the difference of nectar reward between plants of different colors as well as shapes.[9]

We often think of learning as a capacity of dogs or bears or other species familiar to humans, but even among tiny-brained hummingbirds there is the ability to learn. Rather than selection simply for offspring with the "right" hardwired instincts, then, there is probably also selection for the *capacity* to learn.

EVOLUTION IN ACTION: OCEANIC DISPERSAL

Coral reef species are among the widest distributed of all life-forms. In some species, larvae are carried long distances by surface currents. In other species, however, corals travel long distances because adults hitch a ride on floating rafts of kelp. But not all coral-bearing rafts are kelp. One field study at Australia's Great Barrier Reef observed rafts composed of the skeletons of the reef coral (*Symphillia agaricia*) that had washed ashore, dried to a buoyant state, and were then washed back out to sea. These rafts, about the size of a briefcase and weighing about fifteen kilograms (thirty-three pounds) weren't just flotsam, as we might think passing one by in a boat. These were entire floating microhabitats, transporting not only live corals but also a whole community of filamentous algae, goose barnacles, shrimps, oysters, snails, sea slugs, and a multitude of single-celled species.[10] This is a floating ecosystem; its members co-exist and co-evolve.

Who knows what you'll find when you look. I am willing to bet that virtually nothing is known about these fascinating "life-rafts."

There are other rafts as well. One study documented rafts of the giant bladder kelp (*Macrocytis pyrifera*) discovered drifting in the Southern Ocean between South America and Antarctica. Once again, each raft is a small ecosystem composed of one hundred to two hundred kelp plants and their hitchhikers: "colonies" of mussel-like bivalves. And they carry not only adult bivalves but the young as well, showing that kelp rafts can serve as platforms sustaining mollusk populations on long-distance voyages up to thousands of miles.[11] What different selective pressures might these life-forms encounter on such wide and long travels?

Another study showed that colonies of corals may traverse 20,000 to 40,000 kilometers (12,400 miles to 25,000 miles) over several years, such that these corals might make several circuits of the tropical and subtropical Pacific Ocean in their lifetime.[12]

Modes of transport of life-forms include not only their own behavior—crawling, flying, swimming, burrowing, and so on—but also inadvertent transport, as on the legs of seabirds or on floating rafts of what have you. How this shapes the evolution of oceanic species is a question that opens up fifty fascinating doors, any one of which might be a full career of investigation. We still have so much to learn about our own planet that even the erudite "King of the Ants," biologist Edward O. Wilson (b. 1929), in a recent interview at the 2010 National Book Convention, said that his number-one concern now, after a long career, was conservation. Wonderful things are vanishing, right now, and we don't even know what they are. In addition to being willing to bet that almost nothing is known about these oceanic rafts, I'm willing to bet they are threatened by pollution. You can read the transcript of Wilson's interview at http://c-spanvideo.org/program/295631-8.

EVOLUTION IN ACTION: VIRUSES OF THE SEA

More wonders come from the sea, and we learn more about them every day. Rather than fish or even plankton, viruses are the most abundant and genetically diverse "life-forms" in the ocean. I put "life-forms" in quotation marks, because not everyone agrees that viruses are in fact alive. However, they do bear self-replicating molecules of the kind we differentiated from information-free replicators in chapter 2.

How many viruses are there? In a single milliliter (0.03 fluid ounces), there are typically ten million viruses. Sampling so far shows they are more common in shallow water near shores than in deeper and offshore waters, and that they co-occur with bacteria. In total, the combined global weight of viruses is equal to seventy-five million blue whales; about fifteen billion pounds. These oceanic viruses are almost entirely unknown, but initial investigations show they have genomes from 997 base pairs to 1.1 million base pairs in size (recall that the human genome is about 3 billion base pairs long, and most animal genes are about 1,200 base pairs in length[13]). Fascinatingly, comparison of some base-pair sequences showed nearly identical base-pair sequences in viruses taken from Antarctic waters and those of the Gulf of Mexico! And new analysis shows that some viruses have picked up genes from their hosts and spread them among the virus population. Wonders do not cease: on a daily basis, oceanic viruses kill 20 to 40 percent of all marine bacteria.[14]

Again, imagine the complexities here and the sheer quantities! What can we learn from these life-forms—or proto-life-forms? They replicate, vary, and are selected upon in astronomical scale. Their evolution is, to put it mildly, of interest. Imagine these life-forms teeming in every drop of water next time you're at the beach. Don't worry; people go in the water all the time, and have for a very long time, without picking up deadly viruses. But just imagine: that's not just water out there—it is nothing less than a living soup; every green or sparkling drop is charged with life.

EVOLUTION IN ACTION: ECHOLOCATION

When we think of the evolution of complex senses, like seeing or smelling, it's easy to become boggled. Such subtle and effective sensors as eyes and tongues are hard to imagine as resulting from the processes of replication, variation, and selection. But recall that just a glance at an eye or a tongue looks at the sensor as it is right now; it's hard to see the history, but we know the history is important; life-forms don't just pop up out of nowhere. Once, my older brother Mark showed me a complex painting he'd made, a large, somewhat abstracted human skull. "How long did that take you?" I asked, and his reply was "About forty years." The same applies to living organisms and their characteristics, like eyes or tongues. Any individual organism, to paraphrase biologist G. C. Williams, can be considered a historical document.

Also recall that it is simply by the appearance of variations and selection upon them—which weeds out fitness-reducing variations and increases fitness-improving variations—that slight but significant improvements accumulate in the genome. This is the world of "the difference that makes a difference." Like compound interest, replication, variation, and selection generate significant effects over time. Yes, there are constraints to selection, and, no, not every single characteristic of a life-form is an adaptation for something right now. But the principles of replication, variation, and selection remain, and they have shaped the enormous diversity and complexity of life-forms over time.

Echolocation is a fascinating example of a complex adaptation. Consider that among bats there are eight main types of *echolocation* (use of echoes to map out the immediate environment). These include brief tongue clicks that emit a sound, which then bounces off something in the environment, like a stalactite in a cave or a flying insect in the open air. The sound that bounces back is picked up by the bat's ears, giving it information about its environment. A recent study has revealed that cave-dwelling bats of the genus *Rousettus* (including the fruit bats) make these clicks by popping the tongue from the floor of the mouth. The sound goes out, and the echo is picked up by their subtly mobile ears, allowing them to adjust their flight to avoid obstacles in the darkness of caves.

But in bats that spend more time in the open air hunting flying insects, the sound sent out is made by a different clicking action of the mouth, a sound that lasts five times as long as the quick pop of the cave bats. That longer signal, it turns out, is better for detecting flying prey than for navigating around obstacles. Different kinds of echolocation, then, have evolved in different selective environments: in caves and in the open air.[15] How did that longer clicking action evolve? It must have originated in some kind of—you guessed it—variation.

Not only have bats evolved echolocation; echolocation has evolved independently in a completely different life-form, that of the toothed whales.

Specifically, over seventy species of toothed whales use echolocation to find and intercept food. A study of Blainville's beaked whales (*Mesoplodon densirostris*) in waters off West Africa's Canary Islands found that echolocation sounds were generated by the whales' forcing pressurized air through a hole in the forehead. Looking closer than we might look when seeing a whale for a few seconds on TV or even from a boat, researchers found that hunting was carried out in specific search, approach, and terminal phases, each using subtly different variations in echolocation. In the terminal phase, for example, the whales use a *location buzz* when they're only a body length (two to five meters, or about ten feet) from the prey target; this buzz and its returning signal give the whales very rapid updates about location, speed, and direction of the target right up to the moment the whale's jaws clamp down on the prey. This "target lock" buzz is distinctively different from the sounds the whales use during the rest of their foraging activities. When they're just scoping out their target environment, spotting potential targets up to 275 meters (900 feet) away, the whales use different sounds, and leave longer spaces between the sounds than in the target-lock phase. This longer interval (around 350 microseconds) is apparently used to process the multitude of echo returns the whale hears during scanning.

These whales are deep divers; they hardly start echolocating until they're below 200 meters (657 feet) deep, but between that depth and even deeper (they often dive below 500 meters, or 1,640 feet—a depth that would crush most submarines), they echolocate continuously, making up to 15,000 clicks per hunting dive.

As much as this study revealed, plenty remains to be known. For example, just before intercepting their prey, these whales roll onto their backs. They also come at the target not straight on but make a last-minute, lightning-quick turn just before impact. Also, the whales are careful about what they eat. They don't just go after the biggest and easiest targets in the "echo cloud"; they prefer squid and fish on the margins of the cloud, and they prefer them at depths between 650 meters (2,100 feet) and 725 meters (2,400 feet), and nobody knows why (yet).[16]

For most of the history of biology, gaining an understanding of this echolocation would have been difficult if not impossible without behavioral studies in the field and, instead, relying on the comparison of the anatomy of these whales to others. Comparative anatomy is a fine pursuit, but today, in addition to the technology for behavioral studies, we have the jaw-dropping capacity to examine the genes of an organism itself—in this case, the genes related to echolocation.

For example, recent studies show that the gene called *prestin* is involved in development of the hair cells in the inner ear directly responsible for hearing acuity in mammals. Remember, a gene is a section of the DNA that directs all kinds of activities related to growth of the organism. Defects in *prestin*, it's been found, are associated with deafness in humans and mice. A recent investigation of *prestin* compared the gene in mammalian species as diverse as pig, human, cow, hippo, gerbil, rat, horse, bottlenose dolphin, and several bat species. *Prestin* was significantly similar in bats and echolocating bottlenose dolphins (*Tursiops truncatus*),[17] and the inference is that the same modifications of *prestin* have occurred independently in bats and toothed whales because they have been separate forms of life for so very long. This may seem incredible, but remember that an animal gene is often around fifteen hundred base pairs in length, and even single changes in these base pairs can have significant effects, as we saw in chapter 3. Is it so difficult to imagine that similar changes occurred in these two different life-forms, considering that both bat and whale fossils go back over fifty million years? That is a long time, and apparently plenty of time for such parallel development to take place. Understanding precisely how the *prestin* gene directs the development of hearing-related structures is currently under way, and one recent paper delves into the actual mechanical properties of the ear's hearing-related hairs and other structures.[18]

EVOLUTION IN ACTION: SEA LION HUNTING, FROG CANNIBALISM, AND LOCUST SWARMING

Let us move from the small scale back to the larger scale and the complexity of behavior. In one study, Steller sea lions (*Eumetopias jubatus*) observed in Alaska's Prince William Sound were found to feed exclusively on herring (*Clupea pallasi*), which are five times less abundant in these waters than another fish, the walleye pollock (*Theragra chalcogramma*). Even in winter, when there is particular food stress, the sea lions avoid the pollock. Why? The reason appears to do with depth and the time of day: space and time. The very abundant pollock live throughout the day at a depth of about 100 meters (328 feet), whereas herring schools come up nearer the surface to spend the night around 15 to 35 meters (about 50 to 115 feet) down. And the sea lions, for reasons unknown, feed at night. To do this, they arrange in groups of up to fifty, which swim abreast, apparently herding the herring into a feeding ground. Stellers can dive to 250 meters (over 800 feet!), so the mystery is, why don't they go after the pollock? For a reason not understood today, it's apparently easier for Stellers to herd the shallower-occurring herring than dive for the pollock.[19]

A quick survey of this ecosystem would report that there is plenty of food and that by avoiding pollock, the sea lions were acting "suboptimally." But a life-form can't just gobble up anything; in this case, time and space variables are involved. At some point, could the sea lions break up into different populations that feed on different fish at different depths? What if a variation arose that made it easier for some Stellers to dive to the pollock depths than it is today? It's possible that some would start to feed deeper, and some would stay up nearer the surface . . . and who knows where that might lead. Reproductive isolation? Speciation? What we see today might be changing before our eyes, but too slowly to see.

Sometimes "feeding" behavior is ambiguous. In one study of a Romanian frog pond, it was found that when the pond dried up substantially, tadpole metamorphosis sped up, and the young hatched earlier. Not only that, but the hatchlings then began to cannibalize other, less-mature tadpoles in the crowded quarters.[20] We might say this is done to reduce the population, but do tadpoles have the consciousness to determine that there are too many frogs in the pond—to realize such a thing and then act out a plan of cannibalism? Does the proliferation of tadpoles reduce the usual food source so that nothing is left to eat but one another? This seems to be a case of behavioral *plasticity*, the ability of the life-form to vary its behavior as the environment changes. We can imagine that when the pond dries, there is selection for tadpoles carrying genes that direct earlier maturation as well as a propensity for cannibalism. Are those same genes selected against in years when the pond does not dry up? If not, why is this kind of fast maturation and cannibalism normally rare? Could this be an entirely instinctive response that ends up with complex results, as in desert locusts who are solitary *until* the backs of their legs are touched—in crowded conditions—which then "switches" them into an entirely different mode of "swarming" behavior?[21] Although plasticity has been defined and outlined, little is known about it, as we'll see in chapter 8.

EVOLUTION IN ACTION: MOUTH HANDEDNESS IN LAKE TANGANYIKA

In Tanzania's Lake Tanganyika, there are many species of cichlid fishes that dine on the scales of other fishes. To do this, they sneak up from behind and dart in to bite off a scale. Interestingly, they don't come in completely side-on, at ninety degrees, to the prey fish; instead, because of a variation in their lower jaw, their mouths open to the side (just as if you decide to speak out of the side of your mouth), they come in alongside the prey fish, and, with their sidewise-placed mouth, make the grab. And more interestingly, there are two variations of this behavior: the mouth can open either to the left or to the right. In an eleven-year study of this phenomenon, Japanese biologist Michio Hori discovered that this variation is genetically controlled, so that a cichlid is born with either a left- or a right-facing mouth. Therefore, there are two major variations of mouth "handedness." In the species Hori studied, the ratio of left- to right-handers remained basically the same, although there were times when there were a few more left-handers, and other times when right-handers were a little more common. Overall, then, the population is *polymorphic* (*poly*, meaning "more than one," and *morph*, meaning "shape"), indicating that there is more than one distinctive variation in the same anatomical trait—in this case, mouth handedness. We've seen this before in the case of *balancing selection*.[22]

One of the most interesting observations of this study was that members of the prey species were quite vigilant and consistently "kept their eye" on either their left or right flank, depending on whether left- or right-handed mouth types were more common in the population! Over time, could the prey species "select for" (though without any kind of consciousness that we know of) left- or right-handed mouths? Could the left- and right-handed mouth forms diverge behaviorally for some reason, leading to reproductive isolation and then speciation? Why not? Those are the kinds of questions a person who knows something about evolution can think about, and I'd argue that they're good food for thought. Who knows where thought will lead. It does not have to lead to science, of course; many artists take inspiration from nature.

EVOLUTION IN ACTION: THE SOCIAL LIVES OF MICROBES

I can't say it better than the authors of a 2007 paper titled "The Social Lives of Microbes":

> Our understanding of the social lives of microbes has been revolutionized over the past 20 years. It used to be assumed that bacteria and other microorganisms lived relatively independent unicellular lives, without the cooperative behaviors that have provoked so much interest in mammals, birds, and insects. However, a rapidly expanding body of research has completely overturned this idea, showing that microbes indulge in a variety of social behaviors involving complex systems of cooperation, communication, and synchronization.[23]

The authors proceed to enumerate a number of astounding aspects of microscopic life.

For example, it's been discovered that from 6 to 10 percent of the genes of the rod-shaped bacterium *Pseudomonus aeruginosa* are controlled by cell-signaling processes, meaning that this microbe can be strongly controlled by cell-to-cell signaling systems; that is, not just chemicals in the bacterium's environment but specific chemical systems related to cell-to-cell information. Is this communication in microbes? Can we define communication in microbes? Many animal species, including humans, communicate by nonverbal methods, such as chemical scent, and entire ant colonies are orchestrated and organized by chemical communication. The evidence develops.

Other researchers have found that microbes can indeed cooperate—inasmuch as ants, for example, cooperate—in complex actions. For example, microbes can gather and build biofilms, structures that provide shelter and other benefits for the larger population.[24] In one of the more thrilling examples, some microbes assemble into bag-like structures that house populations of microbes. In one case, microbes land on a hard surface and attach there, after which others build on them, creating a film that protects a colony. Later, members of the colony exit the shell and move on, dispersing to new locations and starting the cycle all over again. Another, perhaps even more mind-boggling, example is that of the fruiting body. In the case of some microbes common in soil, when the microbes are short of food they aggregate into a great body, called a *slug*, which then moves in coordination up through the soil. Emerging at the surface, the microbes differentiate, some forming a climbing stalk. Those that form the climbing stalk die without having their own offspring. Those placed at the tip of the stalk break away and float off as spores, landing in new realms where there may be more nutrients. And there is even a genetic component to these behaviors, just as there are genetically controlled behaviors in other life-forms. In microbes that form fruiting bodies, those with a mutation of the *csa* gene—we don't need to know more about that here—build less-effective fruiting bodies than those with the normal form of the gene.

Both of these examples demand as much consideration of social organization and communication as the behavior of termites, ants, and bees, all of whose social lives have been studied intensely.

Yet another discovery is that some microbes produce *common goods*, substances such as adhesives that aid in the metabolism (digestion) of certain foods; but in these populations there are some members, referred to as *cheaters*, that use these products without producing any of their own.

If social activity in microbes doesn't interest you, maybe locomotion—how microbes get around—does. A recent review shows that bacteria don't just float around aimlessly; they move by *swarming*, *twitching*, *social gliding*, *adventurous gliding*, *sliding*, and *spreading*. And this is just the diversity of mobility on *surfaces*, not to mention in the free air or in liquids.[25]

I think you get the idea. There is much more going on in the world of microscopic "anonymous blobs" than many of us ever dreamed.

EVOLUTION IN ACTION: CAVE FISHES OF MEXICO

Can species "devolve"? The term is used loosely by some to document human history; for example the "erosion of social values," but does it actually happen in nature? Not really. While a species could of course be subjected to selective pressures similar to those of its ancestors, "running the DNA backward" doesn't make sense. The DNA, as we've seen, accumulates changes that produce different variations over time, and these do accumulate in the genome. But we don't know of any "time stamp" on these changes that could be systematically "read backward," such that the DNA would simply start "working backward." Still, species *can* adapt to changing environments in ways that result in the *loss* of previously useful structures, leading to what looks like "devolution" or "going backward." For example, many of the Mexican tetra fish (*Astyanax mexicanus*) live in open-air water bodies, but cave-exploring biologists have identified twenty-nine populations living deep in caves. In these populations (named *troglomorphs*—*trog* referring to darkness, and *morph* referring to shape), some new variations have arisen, such as new feeding apparatuses. But these populations have also lost, over time, some characteristics present in their open-air fellows.

Dark-dwelling species commonly "lose" their light-sensing organs over time, and this has occurred in the tetra. And remember, it's not just that some new variety of the tetra popped up in the caves (as an evolution denier would have to propose); young troglomorphs are born with cells that begin to develop into eyes, under direction of the fascinating master eye control gene *pax6*, but rather than developing fully, these eye structures atrophy and sink back into their sockets. Eyes are composed of living tissue that needs calories to stay healthy, and here (as in other cases of troglomorphic loss of surface characteristics) the essential argument is that if one can survive without those structures (as one can in a lightless cave), then a mutation (variation) in which that structure is not present—or, in this case, atrophies rather than matures—isn't just *not a problem*, it might *increase* fitness because the body has to procure fewer calories (feed less, essentially) than its fellows. Therefore, genes that don't direct complete eye formation could be selected for and might well spread . . . as has been the case in these Mexican cave fish and many other troglomorphs.[26]

EVOLUTION IN ACTION: SPECIATION IN ACTION

As I mentioned in chapter 5, those at issue with Darwinian evolution often ask to see speciation in nature. Since species are defined—more or less—as reproductively isolated life-forms, if we want to test whether or not two subpopulations of a given population are diverging—

speciating—the way to do so is by attempting to mate members of these two populations. If the crosses are infertile, well, you have clearly different species. But when, in the divergence of two genomes, do you "call it" and say, "X and Y are indeed diverging"? That is a tough question, and, I'd argue, unnecessarily tough. If X and Y can't mate, we can call it and say these are different species; but if they can mate and have offspring, but the offspring are somehow abnormal, X and Y are diverging. But who's to make the call that 80 percent incompatibility or 51 percent incompatibility "is speciation"? On an important level, the precise figure hardly matters. There will normally be a spectrum, rather than distinctive breaks, and that spectrum could persist for centuries before complete divergence, or speciation. Of course, we always want details, but we also have to be practical. How many breeding experiments can we do? Still, new methods of rapid genome-comparison analysis can help if we want to compare X and Y to search for incipient reproductive isolation between two populations of the same kind of life-form. Such "genomic" studies will revolutionize the study of speciation.

In a recent study of divergence between sea lion populations of the Galapagos Islands, a German, Alaskan, Ecuadorian, and Irish team of biologists (camp evenings must be lively!) compared various characteristics of two populations of sea lions. One population, from the central islands, fed in shallower water, while those from the northwestern island fed more commonly in deeper water. While they were the same species and could, genetically, interbreed, these populations do not breed with one another, even though they live only 200 kilometers (about 120 miles) apart, a distance well within their usual foraging range. Not only do the sea lions differ in some nuances of cranial (head) anatomy, genetically they were found to be very closely related, but not as close to what we would expect of two populations of the same species. The inference, then, is that these groups have become reproductively isolated, for whatever reason, and if things continue in this way, there may well be speciation: two different kinds of sea lion where once there was just one.[27]

AND THE STUDY GOES ON

To give you an idea of just how fascinating is the diversity and detail of current studies in evolution, I've listed below some titles of articles in the 2008 *Annual Review of Ecology, Evolution, and Systematics*. Having read this book, I think you'll find some of these topics reasonably comprehensible. So much to read, so little time!

- "Wake Up and Smell the Roses: The Evolution of Floral Scent"
- "Revisiting the Impact of Inversions in Evolution: From Population Genetic Markers to Drivers of Adaptive Shifts and Speciation?"
- "Herbivory from Individuals to Ecosystems"
- "The Impact of Natural Selection on the Genome: Emerging Patterns in *Drosophila* and *Arabidopsis*"
- "Adaptation to Marginal Habitats"
- "The Evolution of Animal Weapons"
- "Evolutionary Ecology of Figs and Their Associates: Recent Progress and Outstanding Puzzles"
- "The Earliest Land Plants"

Keep in mind that the *Annual Review of Ecology, Evolution, and Systematics* is one of *several thousand* scientific periodicals carrying research on the life sciences. This is a partial list of the contents of a single volume! Imagine just how much is being learned worldwide, year by year. And again, if evolution were easily dismissed, fundamentally flawed, or a "weak" explanation for so many facts, wouldn't somebody notice—and leap at the chance for eternal fame as the person who discovered that all of biology is wrong? But no, the principles of Darwinian evolution as the cumulative result of replication, variation, and selection are reinforced by these studies, every day, all around the globe, at scales from the molecular (DNA) to organism, population, species, and even ecosystem. Evolution is a fact.

EVOLUTION, ACTION, HISTORY, AND COMPLEXITY

We can see now the complexity of living systems, how life scientists learn about them, and, finally, just how much there is to learn. We know plenty, but we are also ignorant of plenty. And yet none of these or other countless studies concludes that the principles of Darwinian evolution are basically flawed. If complex new structures or entire life-forms were found to pop up out of nowhere, then, indeed, Darwinian evolution would have to be put aside. Darwin recognized this: "[I]f it could be demonstrated that any complex organ existed which could not possibly have been formed by numerous, successive, slight modifications, my theory would absolutely break down."[28]

A glance at complexities like echolocation could indeed give rise to the illusion that such things must have been "built" with intent rather than evolving naturally over evolutionary time—a topic I'll return to in chapter 9. But, as we have seen, a glance captures only a moment in time. A glance does not appreciate the history of a painter or the history of an organism. But painters are not born knowing how to paint masterpieces, and organisms with their complexities do not materialize out of nothingness. There is such a thing as history.

Understanding the significance of that history is central to understanding evolution. Occasionally it is claimed that extremely complex structures could not have arisen by replication, variation, and selection, and that in fact the "argument from design" is one of the oldest antievolutionary critiques. The argument dates back to the eighteenth century, when James Paley (1743–1805) proposed that in the same way a timepiece appeared too complex to have evolved, complex life systems, like eyes, also appeared too complex to have evolved and must also have been invented (like the timepiece) by a creator. Writing decades before geologists conclusively demonstrated the ancientness of the earth (among other things),[29] Paley was unaware of the time depth behind an eye, so on the one hand, it's understandable that Paley could not fathom Darwinian evolution.[30] He lacked an understanding of the actual age of the earth, so he simply could not fathom how complex living things could occur other than by instant creation. But on the other hand, not everyone was so limited in their ideas: Roman philosopher Lucretius (born around 90 BCE) believed that the earth was relatively young but *also* that life was the result of natural rather than supernatural processes. After demonstrating that living things, among all others, are composed of small particles (atoms), Lucretius concluded: "At this stage you must admit that whatever is seen to be sentient (living) is nevertheless composed of atoms that are insentient. The phenomena open to our observation do not contradict this conclusion. Rather they lead us by the hand and compel us to believe that the animate is born, as I maintain, of the insentient."[31]

So, even Lucretius, who believed in an earth only a few thousand years old, thought that life was not supernatural in nature and that natural processes could indeed produce complexity. But Lucretius was a heretic,[32] and his ideas were not widely circulated. The vitalist concept that complexity could arise only from a creator flourished right up until the time of Darwin.[33]

Recently, for example, biochemist Michael Behe (b. 1952) of Lehigh University has suggested that the bacterial flagellum, a complex tail-like structure, is *irreducibly complex* (Behe's widely used term) and also that it could not arise simply by evolutionary principles. This is simply Paley's concept, but using a bacterium rather than a timepiece as its example. Evolutionary philosopher Eliot Sober points out that Behe is ignoring the critical fact of evolutionary history:

> [Behe] . . . defines irreducible complexity in terms of the function a structure has now and whether the structure would be able to perform that same function if one of its parts were excised [today]. Behe thinks it is a problem for evolutionary biology if wings now have the same function of promoting flight [as in the past] but would not be able to perform that function if even one of their parts were removed [today]. This is a problem only if flying was the name of the game all along.[34]

In other words, if we think of the complexity of a bacterial flagellum today and how it operates as a mechanism of propulsion today, we might conclude that its complex mechanisms must all have been assembled at the same time for the function that we see today. But this misses the possibility that the flagellum could have developed in small steps *not for what we see today* but for some other functions or for functions in the past. The flagellum, if we consider evolutionary history, did not pop up, fully formed; it was built over time by replication, variation, and selection.[35]

The old, incredulous arguments that complexity could not occur by natural processes are just that: old, propagated by unexamined tradition, uninformed by new discoveries, and incredulous, disbelieving in a past. Of course, echolocation, or the eye, will not pop up wholly formed. Nothing does. But over time, replication, variation, and selection can result in complexity entirely without intent.

This chapter has given us a look at that history *in action*. Who knows what will develop in the future from processes that we see a glimpse of today? To understand, we have to look carefully. It's been suggested before, by Aristotle: "We must not recoil . . . from examining the humbler animals. In all things of nature there is something of the marvelous."[36]

The next chapter shows that we do look carefully indeed, and what marvellous things we find when we do; and in the final chapter I explore why, despite overwhelming evidence, evolution can still steem counterintuitive.

CHAPTER 8

THE MIRROR-HOUSE OF EVOLUTION

Science is an endless search for truth. Any representation of reality we develop can only be partial. There is no finality, sometimes no single best representation. There is only deeper understanding, more revealing and enveloping representations.

—Carl Woese
"A New Biology for a New Century"[1]

Throughout this book I've mentioned that there are many new discoveries as well as plenty of details behind the elements of replication, variation, and selection. In this chapter I'd like to introduce you to some of these discoveries and details. They result from new techniques that have led to new observations and new theories forwarded to explain those new observations. Two things remain clear after reviewing these details. First, they don't undermine basic evolution as Darwin understood it; they support it. Second, just because something cannot be explained today does not mean that it will not be explained in the future—when we know more—or that the default explanation should be supernatural. Scientific explanations grow like plants, rooted in the soil of accumulated knowledge.

So we will find in this chapter that biology is not dropping Darwinian evolution. Replication is found to happen in more ways than we expected;

there are more sources of, and constraints upon, variation than ever dreamed of; and selection operates in different ways on different life-forms at different times. But these three principles remain the backbone of evolution. Our understanding of evolution is being refined, not scrapped. Evolution remains, but our understanding of its complexity is growing.

The modern research literature on evolutionary biology reveals a breathtaking interconnection of observations, concepts, facts, and theories. This is the mirror-house of evolution. Step inside . . .

A NEW EVOLUTIONARY SYNTHESIS

There is so much complexity and detail revealed by new knowledge that major issues in the life sciences are being reconsidered, such that many agree that a "new biology" is in view, or that we're already in it. The modern world of biology is largely shaped by what's referred to as the "modern evolutionary synthesis," which crystallized in the 1940s,[2] when genetics, population genetics, and other elements of older conceptions of evolution became integrated; but dozens of advances and revelations have led many in biology to call for a "new synthesis." Only last year, biologist Eugene V. Koonin titled a paper "*The Origin* at 150: Is a New Evolutionary Synthesis in Sight?" ("Origin" referring to Darwin's *On the Origin of Species*, published in 1859). In that paper, Koonin argues that in the "post-genomic" biology—the biology informed by so many advances in understanding DNA, the genome—while Darwinian evolution still occurs, it is now understood to be a result of a "plurality of evolutionary processes and patterns," resulting in an "incomparably more complex" understanding of the world of living things than ever before.[3] Koonin isn't alone: a 2007 paper was titled "The New Biology: Beyond the Modern Synthesis," and even in 2004, Carl Woese wrote a paper titled "A New Biology for a New Century." The rest of this chapter shows why so many are calling for this "new biology."[4]

MOLECULAR GENETICS

Scientific review papers are written as surveys of the current state of certain topics; they make a place for scientists to sit back and compose an overview of what's going on in their field. These papers are a way to see the forest despite the trees. Nearly every review paper I've read in the research for this book—and there have been about twenty (in addition to several hundred research reports)—begins with a statement about the significant new understanding of biology in the last decade or two. Much of that new understanding comes from the field of molecular genetics: understanding evolution at the level of the genes, which, as we've seen throughout this book, are at the heart of living things and are so crucial to our understanding of the diversity of life. But molecular genetics hasn't become entirely focused on the microscopic, and in fact it has enormously increased our understanding of evolution at every level, as mentioned in a recent review: "Biologists are now addressing the evolution of genetic systems using more than the concepts of population biology alone, and the problems of cell biology using more than the tools of biochemistry and molecular biology alone. It is becoming increasingly clear that solutions to such basic problems as aging, sex, development, and genome size potentially involve elements of biological science at every level of organization, from molecule to population."[5]

Clarifying the significant results of many lines of new investigation resulting from molecular biology, biologist Carl Woese even highlights that while replication, variation, and selection remain the backbone of evolution, "The creation of the enormous amount and degree of novel complexity needed to bring forth modern cells is by no means a matter of waving the usual wand of variation and selection."[6]

So just as the invention of the microscope allowed us to perceive the microbial world, and just as the telescope opened up the world of the very distant, the invention of technologies to directly investigate the genome of any species proves to be revolutionary.

Molecular biology allows us to identify and directly observe (among other things):

- The chemical origins of variation
- The variations among individuals of a single life-form
- The variations between different species
- The date at which life-forms shared a common ancestor

Rapid "gene sequencing" allows us to compare the genomes (entire gene sets specific to individual life-forms) of different life-forms, be they individuals of the same species or individuals of different species.

We've seen already that understanding the origins of variation in base-pair differences between individuals gives us great insight into the nuts and bolts of evolution. Remember, just a single base-pair difference causes different foraging behavior in flatworms, different alcohol-processing capability in fruit flies, and (though I haven't mentioned it yet), in humans, it determines whether or not one has sickle-cell anemia.

Also, comparing the genes of different individuals of a species can show just how much variation there is within a population, and what kind; it might even identify *cryptic* variation, or variation that exists on the genetic level but is otherwise invisible.

And what about comparing the DNA of members of different species, like comparing fingerprints? Not only does it allow us to see how "close" one species is to another (as we saw in the case of the fishes of the

Nicaraguan volcanic lake), but we can use this new knowledge to—completely independently—"check" the millions of species labels that have been made over the last few centuries—labels that were built on anatomical differences. The new field of *comparative genomics* is under way, and it is thrilling to imagine what we might learn from it. Recently the Woods Hole Oceanographic Institution's deep submersible *Nereus* "landed" at 10,900 meters (6.7 miles!) below the surface of the western Pacific. Within minutes, the crew at the surface saw through video monitors a small seafloor creature making its way across the mud. You can bet that before long, the genes of this creature will be compared with life-forms that live much nearer the surface. Are they completely different? Are they related?

Molecular genetics also allow us to date the divergence of species. Generally speaking, a certain kind of DNA accumulates mutations (variations) at a known and stable rate. If a population splits into, say, new populations A and B, in X generations there should be X mutations (differences) between the genomes of populations A and B. Now, get a biologist to compare the genomes of A and B, and you can count up the differences between them. The number of these differences is your "clock," telling you how long A and B have been separate. In one of the most spectacular cases, the genetic clock has been used to date the emergence of modern humans from Africa, somewhere around 150,000 years ago.[7] And again, this is independent evidence that largely corroborates not only the fossil evidence (which suggests the same thing) but also the archaeological evidence (which suggests the same thing). The molecular clock has its complications, but it's being refined rather than abandoned. It's hard to argue with such compelling evidence from multiple and completely independent studies. This is the kind of evidence we could take to court, and in fact DNA evidence is used in courts worldwide, every day, and it is all based on the principles of evolution by replication, variation, and selection.

But we need to be careful. It's easy to think that a species' "essence" has been captured by sequencing its DNA, but because every individual differs, a recent review cautioned that "a single genome sequence obtained in a laboratory should not be treated as an endpoint in an organism's evolution, nor is it necessarily representative."[8]

For this reason, genomes "known" for a given species are considered "reference" genomes, with the understanding that there will be variations among members of that species. This doesn't undermine the use of DNA evidence in court, however, because that is usually used to identify *individuals* of the species *Homo sapiens sapiens*, not whether or not they're in different species.

In short, viewing DNA with genomic tools allows us to see the nuts and bolts of evolution; we have opened the hood and are now in with the engine, able to look at the fuel pump, pull out the spark plugs, and so on. The effects have been revolutionary. Not all the recent advances are related to genomics, but many are.

HORIZONTAL GENE TRANSFER (HGT)

Back in chapters 2 and 3, we saw that life-forms come from parent generations and that their resemblance to their parents was a result of the transfer of DNA from one generation to the next; this is referred to as *vertical gene transfer*. While this is true, it turns out that there is another mechanism by which some life-forms "get" their DNA. This is called *horizontal gene transfer* (HGT); *horizontal* (or, sometimes, *lateral*) referring to the fact that DNA can sometimes be picked up during the course of life and incorporated into a life-form's own DNA and passed on to the offspring. For instance, 18 percent of the genome of *Escherichia coli* appears to have derived from another well-known bacterium, *Salmonella*.

It's been known since the 1930s that this happens among microbes, but only in the past decade—with advances in genomics that allow us to compare entire genomes—has the extent and significance of HGT been appreciated. Just ten years ago, marine biologist J. H. Paul wrote: "Not too long ago, attributing an unusual genomic finding to a lateral transfer event would have been met with general skepticism. . . . However, complete genome analysis has made the concept of gene transfer actually indispensable to understanding microbial evolution."[9]

One of the ramifications of HGT is that it makes the classification of these life-forms difficult; remember, a species is a reproductively isolated group; but HGT allows "mobile" DNA fragments to float freely, being picked up by different "species" of microbes, making it very hard to say this "species" of bacteria is different from that "species."[10] Bacteria, microbiologists have recently written, are "constantly bombarded with foreign genes,"[11] and they get their DNA not just from their ancestors but also from outside sources—even from sources very different from their own kingdom of living things. Horizontal gene transfer, it seems, isn't an exception, a one-off; it's "pervasive," often imparting significant amounts of new variation to bacterial life-forms.[12] Since microbes reproduce asexually, without the DNA "shuffling" ("recombination," I called it in chapter 2) that supplies so much variation in sexually replicating life-forms, many have for a long time wondered how microbes have lasted so long; what is the source of variation that they need to survive in ever-changing conditions? A lot, it seems, comes from horizontal gene transfer, leading HGT to be considered a critical factor in the evolution of microbial life: "Scientific opinion has now shifted and favors a significant role for horizontal gene transfer in prokaryotic genome evolution."[13]

Why does the microbial world matter? One reason is that, although invisible to the naked eye, microscopic life is actually the most common on Earth: there are uncountable numbers of these life-forms, and if evolution works among them in ways different from macroscopic (normally multicellular) life-forms, and we want to understand evolution, we can't ignore the most common life-forms on the planet! Also, early life on Earth must have been unicellular, with many parallels to unicellular, microbial life today, so understanding HGT can help us understand the evolution of early Earth life; it may well be that billions of years ago, biology was rather different than it is today, as suggested by biologists G. J. Olsen and Carl Woese: "The evolutionary picture emerging from genomics, as we trace our ancestry further and further back into the past, is an unsettling, yet entrancing one. Looking back towards the Universal Ancestor, the simple world of distinct organismal lineages ["types," for example, species] and a robust, well-defined connection between the genotype and phenotype—foundation stones of biology—loses its substance, dissolving in the turbulent evolutionary dynamic that shaped genotype, phenotype, and their connection."[14]

Exploring that ancient world, biologist Carl Woese has proposed that indeed, early in the evolution of life "HGT was the primary force in early cellular evolution, predominating over the vertical inheritance of genes, because early cells were extremely simple, loosely organized and communal. These early cells had highly error-prone DNA replication and expression systems, which facilitated the integration of DNA from other cells in the ecosystem. As cells became more structured, a 'Darwinian Threshold' was reached, at which vertical inheritance predominated as the main path of gene flow."[15]

So far I've said that HGT is prevalent among microbes, but a few instances of it have been observed in multicellular life as well. We may well find—and some argue that we already have found—that HGT has played and continues to play an important part in the evolution of multicellular life, as marveled at by biologist T. R. E. Southwood: "Incredibly we come to realize that multicellular organisms are not strictly the end products of a monophyletic [single line] of DNA, but really complex organisms that have collected bits and pieces from different lines of the evolutionary road."[16]

Among some flies, for example, those species we know so well from so many genetic studies, HGT has been observed, and it's also known to occur in plants. Biologists are now classifying the different kinds of HGT. One recent paper distinguishes, for example, between HGT that happened long ago in a species' history and HGT that has occurred more recently.[17] That's basic work, but each has its own implications, and this sketching out of basics is the necessary, exploratory, descriptive work of any new branch of science. Life scientists are also scouring the various genomes so far "decoded," searching for evidence of ancient HGT events.

Perhaps the most surprising result of recognizing the scale and significance of HGT is that it shows that a kind of "Lamarckian" evolution does indeed occur. Jean-Baptiste Lamarck (mentioned in the notes to chapter 3) was a French naturalist who, before Darwin, proposed that species could change through time (this is why we distinguish between *Darwinian* and other kinds of evolution). But in contrast to Darwin, Lamarck proposed that individuals would change in ways significant to their offspring as a result of things that happened to them during the course of life. However, Darwinian evolution proved Lamarckism to be essentially false. As I mentioned in chapter 3, if I lose an arm in an accident, that "information"—acquired during the course of my life—does not encode somehow in my genes, such that my offspring are born missing an arm. That remains clear, but the revelations of HGT suggest that *some* information, in the form of DNA, can be acquired by lifeforms during the course of life and passed on to the offspring: a quasi-Lamarckian *inheritance of acquired characteristics*.[18]

Both molecular biology and the revelations of HGT are even

prompting some to rethink how we classify and depict our classification of life-forms. Figure 8-1 shows a number of ways to classify living things. In (A) we see the archaic "pigeonhole" concept in which a hard line is easily drawn around kinds of life; this also shows these life-forms arranged in an hierarchy once known as the Great Chain of Being, which maintained humanity near the top (not shown here are the supernatural beings that would be above even humanity). In (B) we see that time (PRESENT and PAST) has been included, recognizing that life-forms aren't the unchanging, fixed types we saw in A, but rather they change through time. On the left of panel (B), we see a conception in which a life-form changes gradually over time, and if we find remains of this life-form, we might label them A and B with dotted rather than hard lines, recognizing that every individual in a lineage is a "missing link" between past and more recent kinds. On the right side of (B), we see that species A changes over time into F, G, and H, indicated by overlapping boxes that reflect kinds of life. A also gives rise to B, however, which is an ancestor of C, D, and E. In panel (C) of this figure, we see two methods of showing the relationships of certain life-forms (for example, A, B, C, D, and E across the top). Again, the present is at the top, and the past is at the bottom. A glance at the left of this diagram shows that E is quite a loner, different from the rest, whereas C and D are closely related; they diverged relatively recently and are somewhat related to B. On the right, the same diagram shows how the revelations of HGT are being incorporated into such "trees." Here we see that at some point in the past (marked by an arrow) some genetic material from lineage E "jumped" across lineages and became incorporated into lineage B. In the same way, a little more recently some DNA "jumped" from C and was incorporated into lineage E.[19]

The recognition of HGT has been so significant to understanding evolution that in 2009 biologist Eugene V. Koonin titled a paper "Is

Fig. 8-1. Ways of Showing the Relationships between Living Things

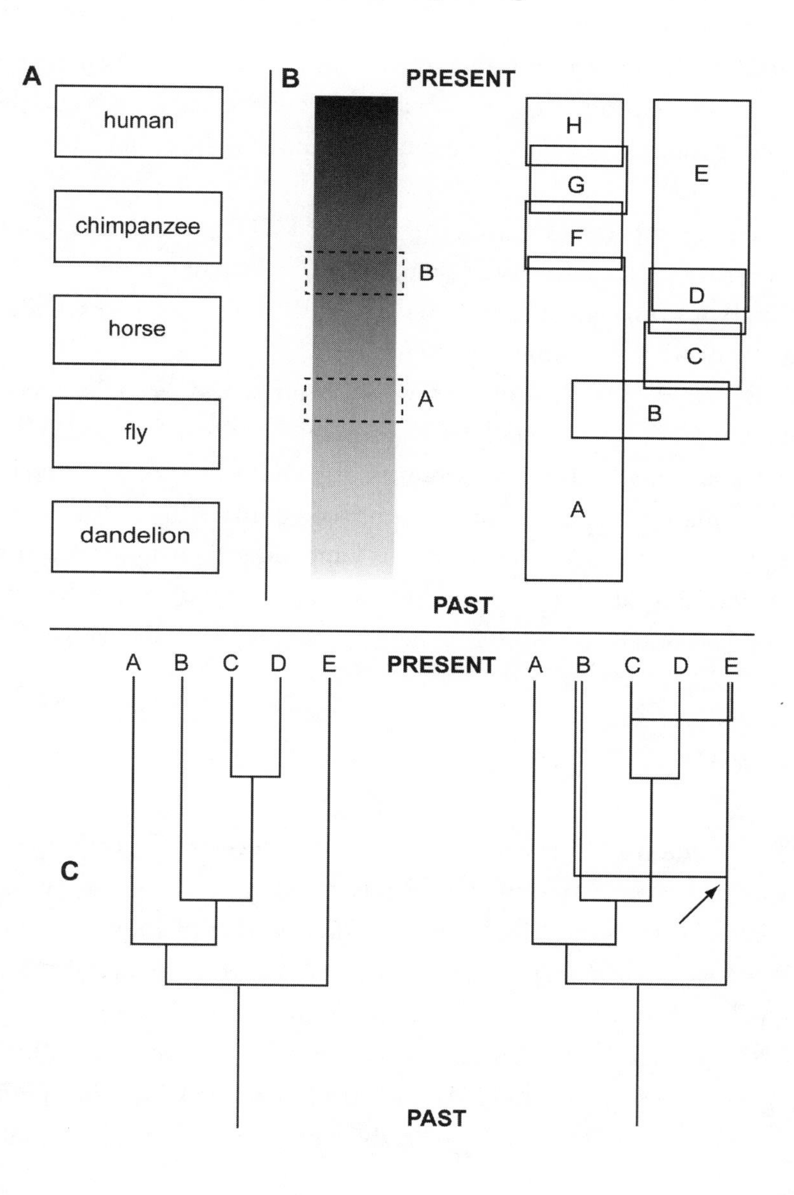

Evolution Darwinian or/and Lamarckian?"[20] The startling answer to this startling question seems to be *both*. And worldwide, work is well under way to investigate this phenomenon. A 2009 book titled *Horizontal Gene Transfer: Genomes in Flux* includes a chapter titled "Defining the Mobilome," which sketches out the basic kinds and properties of "mobile genetic elements" that can move from one kind of life to another.[21]

We can expect more from the study of HGT. As recently as 2006 it was written that "the sheer scope of the effect of mobile DNA on prokaryote [unicellular] genomes has only come into focus since the advent of whole-genome sequencing."[22]

By 2009, several thousand viral genomes had been "sequenced" (their every A, C, G, and T base-pair sequence identified and archived for research), as well as the genomes of about a thousand bacteria and their relatives and about a hundred multicellular life-forms.[23] Sequencing of many other species is happening right now, as you read this, in many labs across the world. Who knows what genes have been moved around by horizontal rather than vertical movement?

PHENOTYPIC PLASTICITY

The phenotype—the body built by the genes—is not entirely set in stone in the sense that once it begins growing, nothing else can occur except growth. So far I've focused on the fact that phenotypes are built by genotypes, and that is true and fine. But during the life of the individual organism, it can change in important ways. *Phenotypic plasticity* (*plasticity*, referring to changeability) refers to the fact that individual organisms can change in physiology (bodily processes) or morphology (shape) in response to changes in the environment.[24] It's important to remember that this isn't just an organism's usual physiological reaction to the environment, such as when an animal gets larger, for example, as it eats more food. In phenotypic plasticity, *specific genes directly control the changes in the body in reaction to environmental cues.* These specific genes, then, are specifically "activated" under certain conditions. Since they "live" on the DNA, these genes can be selected for (if they increase fitness) or against (if they decrease fitness).

One ramification of phenotypic plasticity is that when observing the characteristics of an organism, it's important to identify whether what you're looking at (the size of a crab's claw, the coloration of a fish) is simply growth (for example) or a genetically directed plastic reaction.

For example, the water flea (genus *Daphnia*), a tiny, shrimp-like crustacean, normally has a teardrop shape, with a long, thin tail at the end of a globular body. But when the water flea is exposed to certain chemicals floating in the water—chemicals associated with predators dangerous to the water flea, such as the phantom midge (*Chaoborus americanus*)—the water flea grows a long, thornlike spike on the front end of the globular body. This spike, which provides some protection from predation, is nearly the size of the body itself, so it's no minor characteristic, and it "comes on" only when the predator's chemical scent is detected. Because such specific body structures are built by the DNA, somehow the chemical environment activates certain genes that direct the growth of the spike; this is the activation of a *switch*.[25] And this means that water fleas carry a gene or genes that lie dormant until the signals of predatory fish are detected, and this gene or genes are passed on to the next generation during reproduction.[26] Not only is there selection, then, on the phenotype (body) born into an environment, *but there must be selection on genes that direct plastic reactions*. Biologist Mary Jane West-Eberhard indicates the significance of plasticity this way: "Genetic variation and developmental plasticity are fundamental properties of all living things; all individual organisms, with the exception of mutation-free clones, have distinctive genomes, and all of them have phenotypes that respond to genomic and environmental inputs."[27]

Keep in mind, also, that not just any variation can come up as a plastic reaction; there is, as we saw earlier, a genetic history to a species that largely conditions what its genes can and cannot build in the next generation (although, complicating matters, life-forms do of course accumulate new variations over time). Biologists often begin an investigation of phenotypic plasticity by trying to get a handle on the species' *reaction norm*, its usual range of plastic responses.

Phenotypic plasticity has been recognized for some time, particularly among plants, but only in the last decade or so has it really been carefully considered,[28] partly because today we have the genetic tools to examine the specific *sources* of plastic reactions: the genes.

All this allows us a more informed understanding of any life-form. Consider: when we look at a life-form's characteristics—the thickness of its shell, for example, or the broadness of a leaf—are we looking at a "fixed character," something the individual was born bound to develop, or does that character or trait appear the way it does because it's a plastic character that has been "activated" by an environmental cue? Plenty of observations and experiments might be needed to find out. For example, in the case of the water flea, in 1912 biologists thought the ones with defensive spikes were different species from the ones without them. That is worth thinking about!

THE RECOVERY OF ANCIENT DNA

You now know well the significance of DNA. Carrying the instructions—more or less—for building a given life-form, DNA tells us how life-forms are built.[29] For a long time, it was considered that the DNA molecule would be degraded after the death of the body that housed the DNA (the phenotype) in only a few years or centuries. But recent studies reveal that substantial amounts of DNA can be preserved in certain conditions for much, much longer. Hold onto your hat for this next section!

One of the most spectacular examples of ancient DNA recovery comes from Europe, where enough DNA has been recovered from the remains of Neanderthals, proto-humans that survived in Europe between about three hundred thousand years ago and about thirty thousand years ago, when they became extinct. In 2006, a team of researchers based at Germany's Max Planck Institute published a report with an unusually thrilling title: "Analysis of One Million Base Pairs of Neanderthal DNA." And recently the University of California at Santa Cruz has put online the "first draft" of the Neanderthal genome for anyone to browse. Could things get any more exciting? Yes—recently it was found that up to 4 percent of the DNA of modern Europeans is shared with Neanderthals. While it's clear that Neanderthals did become extinct—we don't see them walking around today—this means that the modern humans who emerged from Africa and largely replaced the Neanderthals in Europe must have interbred with Neanderthals to some extent. Things change so quickly; the paper reporting this result came out three days after I finished a class on Neanderthals by telling my students that there's no good genetic evidence that Neanderthals and modern humans significantly interbred![30]

In another case, ancient DNA has been recovered from the nineteen-thousand-year-old eggshells of a number of Australian, New Zealand, and Madagascar birds, allowing researchers to investigate their evolutionary histories in ways undreamed-of even just ten years ago.[31] Some of the oldest DNA so far recovered is from a fossil magnolia leaf in Idaho dated to between seventeen million and twenty million years ago,[32] and in the early 1990s, twenty-five- to thirty-million-year-old DNA was recovered from a termite encased in a glob of amber.[33] And, recently soft tissues—actual blood vessels, for example—were observed in sixty-eight-million-year-old *Tyrannosaurus rex* fossils as well as eighty-million-year-old duckbill dinosaur fossils.[34]

How far back can we go? Recently, Svante Pääbo, an ancient DNA expert at the Max Planck Institute, suggested that a million years is probably the limit in most cases, and that the extremely old DNA I've just mentioned is probably very rare.[35] But how many fossils and globs of amber are there? Millions. Billions! And if there's one thing that has happened consistently in ancient DNA studies, it is that we keep finding that we're wrong about such estimates. With all respect to Dr. Pääbo, I hope he's wrong. Who will be first to recover DNA from the two- to four-million-year-old early hominin fossils—ancestors of our own lineage, *Homo*? I'll bet there's someone writing up the research permit right now . . .

MUTAGENESIS

We know that variation is critical to evolution and that although the common use of the word *mutation* implies something bad or wrong, in evolution it simply means a novel difference between the offspring and its parents and/or its peers. If that mutation affects fitness, we can think of it as "a difference that makes a difference." You know all this by now. But I haven't yet revealed that for a long time it was thought that mutation was relatively rare. Characterizing the concept of mutation in biology over many decades, in 1989 geneticist Evelyn Witkin (b. 1921) wrote that for a long time the prevailing notion "was that mutations were instantaneous events—the mutagen went 'zap'! and that was that."[36]

Things are very different today. Better understanding and direct observation of the genome have completely turned the tables: "The stability of DNA sequences across generations is dependent upon a diverse set of biochemical mechanisms that protect the DNA from predictable sources of damage that otherwise would be catastrophic."[37]

When I first read that passage, I wrote on the margin of the paper "WHAT?!?!"

But it's true. In the bacterium *Escherichia coli*, for example, a gene called *mutT* "sanitizes" DNA by preventing DNA-degrading chemicals from assembling in the first place. Experiments show that if *mutT* is inactive, however, mutations occur at ten times the usual rate. And in humans, a *homologue* (basically a similar gene) has been discovered that does just what the bacteria's *mutT* does; another human gene, called *NER* (*n*ucleotide *e*xcision *r*epair), is active in repairing damage to skin by ultraviolet radiation, which is part of sunlight.

Mutation is not only common, it's so common that a whole slew of DNA-repair mechanisms have evolved. Instead of a "zap" here and there, there is constant mutation and constant repair, such that the whole field has changed from viewing the origins of mutations—variations—from "zaps" to the *failure of DNA-repair mechanisms*![38] These discoveries force us to think of the source of variation—mutation—in a completely new way.

DEVELOPMENTAL EVOLUTIONARY BIOLOGY (EVO-DEVO)

It's been known for a very long time that life-forms develop from DNA instructions, and early biologists were fascinated to observe the development of various life-forms from the embryo to the adult stages. You don't even have to understand what the early twentieth-century biologist Theodor Boveri was writing about to feel his excitement at learning about life-form development in his journal entry from December 1, 1901:

> I have raised the four blastomeres of simultaneous-tetrafoil eggs in isolation. Most important, as a rule a different thing happens to each cell. One goes to pieces at the blastula stage, one forms mesenchyme and then goes *kaputt*, another begins gastrulation or even completes it. Once I even got a pluteus—admittedly somewhat rudimentary. Taking everything into consideration, I believe that here we are finally closing in on the nucleus. That the development does not depend on quantity of chromatin but on quality is quite certain.[39]

And once again, in the past twenty years, advances in genetics have allowed us to watch that development much more closely than with any microscope; in fact, today we can observe and understand the molecular basis—the genes composed of As, Cs, Gs, and Ts—regulating the development of life-forms. Evolutionary-developmental (evo-devo) biology focuses on this fascinating field and has resulted in some truly stunning discoveries.

One of these discoveries is that most animals, "no matter how different in appearance, share several families of genes that regulate major aspects of body pattern."[40]

This means that however different animal life is on the surface—from sea urchin to pine tree to caterpillar to human being—a great deal of that difference is driven by a handful of groups of genes that control the development of certain structures during the formation of the body—the phenotype. Not only that, but the way it works is that if a certain body-building gene is "shut off" at a certain time, a certain kind of body shape is built; whereas if it's "left on"—continuing to direct the assembly of proteins—it can build a very different body structure. The timing schedule of a relatively small set of genes, then, accounts for a large amount of the diversity of animal life.

So, astoundingly, many animal genomes share a "toolbox" of genes that do much the same thing in each animal life-form but result in different bodies—phenotypes—because they have different regulatory schedules. Before having a look at some of these genes, remember that today the naming of genes is a wild and wooly affair; there are standards that result in highly technical gene designations, but life scientists investigating the genes also give newly understood genes memorable and sometimes playful names. The *tinman* genes, for example, are related to the development of the "circulatory pump," otherwise known as the heart. The *sonic hedgehog* genes are significantly involved in a number of organism developments including, specifically, the number of digits (fingers and toes) an animal develops; and certain variations in the *methusela* genes are related to longer life in fruit flies—and humans.

Among the genes common in many animals is the *pax6* gene related to the development of photoreceptors; in some animals these photoreceptors are relatively simple light-detecting structures, but in others, they are entire, complex eyes—but they're all controlled by the "master eye control gene" *pax6*. It is astonishing to find that if a *pax6* gene is transferred from a mammal (like a mouse) to a fly at a certain time of development, the fly will indeed develop photoreceptor structures; and if a fly's *pax6* is transferred to a mammal, it will begin to grow multifaceted, compound fly eyes![41]

Another fascinating example is the *tinman* gene set mentioned above, more specifically *tinman/NK2.5*. Whether in human, fish, fly, or frog (and in many other life-forms), the circulatory pump—the heart—is in part controlled by the same *tinman/NK2.5* gene.[42]

And remember the *prestin* gene from chapter 7? That gene is involved in the development of hearing structures, and in echolocating animals as different as bats and whales, that same gene apparently underwent very similar changes through time leading to similarities in the bat and whale hearing systems. In a similar case, very subtle characteristics of the gene *RH1*, which is involved in the production of *rhodopsin*, a light-sensitive chemical that exists in eyes, have recently been found to be similar in bats that inhabit low-light environments.[43]

It's thrilling to read some of the reports that identify and describe these newly understood genes; the authors write in terms of exploration, traversing "vast realms" of DNA that is currently poorly understood, encountering "dark matter" that is completely unknown, and even finding "hobo genes" that seem to float freely in the world. The whole microscopic DNA domain is nothing less than an unexplored world. Its explorers may as well be surveying the surface of Pluto.

Life scientists have started to identify the main *families* of genes (groups of genes with related functions), even identifying the basic "genetic toolbox" that builds all animal life. The fact that a relatively small set of genes—maybe around five hundred—controls a wide range of basic animal structure developments like eyes and hearing systems has led to the mind-bending concept of "immortal genes." Whatever was the fate of *other* genes in animal life, *these* immortal genes have been preserved for hundreds of millions of years. How else could we explain the fact that putting a mouse's *pax6* gene into a fly would direct the basic assembly of photoreception systems in the fly, considering that flies and mice diverged from a common ancestor hundreds of millions of years ago? We can't. The photoreceptor gene or genes have been *conserved* for eons.

For me, one of the most astounding results of evo-devo is the reconstruction, based on genes rather than the fossil record, of a very early life-form. Using various methods to date the age of certain genes, evo-devo giant Sean B. Carroll of the University of Wisconsin–Madison has observed the distribution of the oldest of these genes across many and widely divergent life-forms, resulting in the reconstruction of an early life-form from which it seems most animal life—including humans—descended.

In figure 8-2 (A), you see my own adaptation of one of Carroll's diagrams of this over two-hundred-million-year-old ancestor. This incredible image reconstructs an ancient life-form from the genes we know it possessed; only a few are mentioned here. We know the animal was segmented from front to back because the *engrailed* gene is so ancient, and we know it possessed a circulatory pump (heart; which implies vessels to carry a kind of blood) because the *tinman/NK2.5* gene is so old, as is the *ems* gene, related to the production of a nervous system, shown here by a dashed "nerve fiber." And there is a gut, indicated by the ancient *ParaHox* genes, and body outgrowths, such as feelers, indicated by the *DII* genes conserved in most life-forms today that have such outgrowths. Finally, a photosensitive eyespot is indicated by the venerable *pax6* gene. I have spent a long time staring at this image, thinking about what it means. In the way that Watson and Crick were "DNA looking in the mirror," this diagram is, in essence, *you* looking at a mirror from vast ages ago.[44] In the lower part of figure 8-2 (B), you see my own reconstruction of a much more recent (but still about sixty-million-year-old) life-form, an early primate. This reconstruction is based on much more traditional methods, such as examining fossils. In this case, the fossils are very tiny, indicating a cat-sized animal with grasping toes and a tail; the woodland camouflage I've put on the coat is entirely conjecture, but it makes sense for the leafy, shadowy habitat we know from other evidence these early primates lived in. I wonder how much I'll be revising this drawing as molecular evidence is used to investigate primate evolution . . .

Evolutionary-developmental biology has reaped a grand harvest. I can't cover it all, but I'd like to give you a sense of just how exciting it is with some titles of recent research reports—you can find any of these papers by searching for their titles on the Internet:

Fig. 8-2. Reconstructions of Ancient Life-Forms

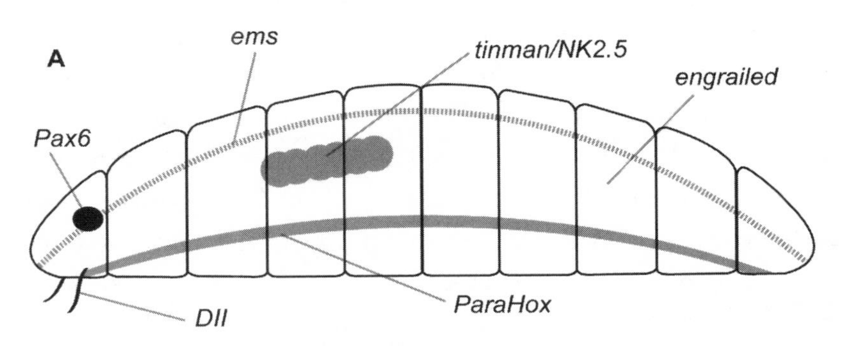

- "Sea Anemone Genome Reveals Ancestral Eumetazoan Gene Repertoire and Genomic Organization" (2007)
- "The Hearing Gene *Prestin* Unites Echolocating Bats and Whales" (2009)
- "The Genetics and Evo-Devo of Butterfly Wing Patterns" (2002)
- "Crustacean Appendage Evolution Associated with Changes in *Hox* Gene Expression" (1997)

Finally, to give you an idea of just what's involved in the identification and comparison of these genes, below I present the actual 2,098 base pairs—the unique sequence of As, Cs, Gs, and Ts of the *prestin* gene—in *Homo sapiens sapiens*. This sequence of information, as hard as it may be to imagine, is encoded in your own DNA, the DNA of every other human being, and—with some variations—in the DNA of many other life-forms. This code governs, in a general way and with other genes (as we saw in chapter 2) the building of hearing structures. Other genes are involved, but *prestin* is critical to hearing. Here we go:

The 2,098 Base Pairs of the *Prestin* Gene in *Homo sapiens sapiens*

acctggaggcagcgcgcgtcgaagaggcagcggctgtggagcgcggcggggcggctcc
gcccagggcagcccgggctgggccaaggagcgagctctcccttctcctgctctcagcctc
agtgatcaaggcttcagtgaactgcactggagctcccagcgggggatcttgtcccctgtc
ccgacttttgtgctgcacattggatctggtgacactcaggaaatgcttgtctccggctgt
taaggaataatttcagagtactatggatcatgctgaagaaaatgaaatccttgcagcaac
ccagaggtactatgtggaaaggcctatctttagtcatccggtcctccaggaaagactaca
cacaaaggacaaggttcctgattccattgcggataagctgaaacaggcattcacatgtac
tcctaaaaaaataagaaatatcatttatatgttcctacccataactaaatggctgccagc
atacaaattcaaggaatatgtgttgggtgacttggtctcaggcataagcacaggggtgct

tcagcttcctcaaggtccttttgctgttattagcctgatgattggtggtgtagctgttcg
attagtaccagatgatatagtcattccaggaggagtaaatgcaaccaatggcacagaggc
cagagatgccttgagagtgaaagtcgccatgtctgtgaccttactttcaggaatcattca
gttttgcctaggtgtctgtaggtttggatttgtggccatatatctcacagagcctctggt
ccgtgggtttaccaccgcagcagctgtgcatgtcttcacctccatgttaaaatatctgtt
tggagttaaaacaaagcggtacagtggaatcttttccgtggtgtatgcgtcgggctgatg
gtttttggtttgctgttgggtggcaaggagtttaatgagagatttaaagagaaattgccg
gcgcctattcctttagagttctttgcggtcgtaatgggaactggcatttcagctgggttt
aacttgaaagaatcatacaatgtggatgtcgttggaacacttcctctagggctgctacct
ccagccaatccggacaccagcctcttccaccttgtgtacgtagatgccattgccatagcc
atcgttggattttcagtgaccatctccatggccaagaccttagcaaataaacatggctac
caggttgacggcaatcaggagctcattgccctgggactgtgcaattccattggctcactc
ttccagaccttttcaatttcatgctccttgtctcgaagccttgttcaggagggaaccggt
gggaagacacagcttgcaggttgtttggcctcattaatgattctgctggtcatattagca
actggattcctctttgaatcattgccccaggctgtgctgtcggccattgtgattgtcaac
ctgaagggaatgtttatgcagttctcagatctcccctttttctggagaaccagcaaaata
gagctgaccatctggcttaccatttttgtgtcctccttgttcctgggattggactatggt
ttgatcactgctgtgatcattgctctgctgactgtgatttacagaacacagaggtgagtg
cccagattggaatgggtgtgaatgtcccggcagagatgacaatgttgactttaggtgtag
accaaagtttaagttggtagaagtggagccctttgatgatttctagttagcgtgagaggg
agctataacactcatgtagcctgttgactagatgaacaaaatgccaatttaaaaattcca
tataattttgccaaatgctcttctatgtcacaatttatgctcccatcaatggttatgtta
aaagagcctaatttccatcattgtttctgccattcctggtctagtgctatgctggtttat
ttatcctcttgtgatttgttttggcaccaagtactgacatgagcttcaatgacatgaagc
aaactctgacaccaagttatcgtatgcattccttccactgtcatttcctccaccctgaac
cactttcccttgttatctcttctccctagtgggaagctgagcccactagggaaagtat

Life scientists have only begun to chart out the properties and effects of genes like *prestin*.

While the discoveries of evo-devo are so fantastically interesting—again, it allows us to see the very instructions for the building of various life-form characteristics—they are very new, and this novelty commands a lot of public attention. Evo-devo, it's sometimes argued, is a spectacle, but not at the heart of evolution; biologists have recognized the significance of development for a long time, and the popular excitement about it is too much. I could not agree less. The revelations of evolutionary-developmental biology *are* a spectacle, they *are* fantastically exciting, and they *will* help us to understand our ancestors and evolution at large. Even from the heart of the most staid, traditional evolutionary studies—the study of fossil life-forms—comes a statement from a leader in the field (Simon C. Morris), recognizing the astounding fact of the "immortal genes": "Spectacularly, many key developmental genes and gene families are shared between all animals. Comparative developmental genetics has shown us that while morphologically [in terms of shape] the animal phyla [major groups] might be 'apples and oranges' they are fundamentally comparable."[45]

If sometimes the results of evo-devo are a little sensationalized in the popular press, well, I can live with that. Evolutionary-developmental biology isn't the rewriting of biology, and it's not all of biology, but what it's showing us about evolution is very important.

MUTUALISMS AND CO-EVOLUTION

Life scientists have always known that no life-form exists in a vacuum; all species have important interactions with other life-forms. For example, a zebra's world is significantly conditioned by the properties of the grasses it eats; and the grass's world is significantly conditioned by other grass eaters, parasites, weather, and so on. Every life-form lives in relationships with other life-forms, and their interactions—whether parasitic (one benefits at the other's expense), symbiotic (both benefit), or what have you—are important. In an article titled "The Evolution of Species Interactions," University of California–Santa Cruz ecologist

John N. Thompson defines co-evolution as "the process by which species undergo reciprocal evolutionary change" and indicates just how important it is:

> Most living organisms have evolved in ways that absolutely require them to use a combination of their own genetic machinery and that of one or more other species if they are to survive and reproduce. . . . We now have convincing examples of coevolution (a) forging obligate mutualisms among free-living species such as yucca and yucca moths, (b) creating divergence in traits among competing fish, lizards, mammals and other taxa [life-forms], (c) producing locally-matched chemical defenses in plants and counterdefenses in insects and (d) maintaining genetic diversity among populations of parasites and hosts.[46]

There is plenty to think about, then, when you see a bird at the park: with what species does it co-evolve?

In figure 8-3, you see the case of hermit crabs existing in close connection with sea anemones: in some hermit crab species, the crab plucks young anemones from the seafloor and places them on their own bellies; the anemones attach and are able to feed on scraps that float free when the crab eats. The crab, on the other hand, receives a protective shell in the form of the anemone's thick, fleshy "foot," which grows up and over the hermit crab's otherwise-unprotected back (A). These life-forms do not just evolve, they *co-evolve.*

In another example, we humans co-exist with about two thousand species of bacteria that thrive in our guts and in many other places throughout our bodies; and squid that use light to avoid predation, and to attract mates, live in symbiosis with the bioluminescent microbes that the squid absorbs into its skin during the course of life![47]

Finally (though there are many more examples), among the life-forms that eat wood—such as termites—a significant component of

Fig. 8-3. Symbiosis

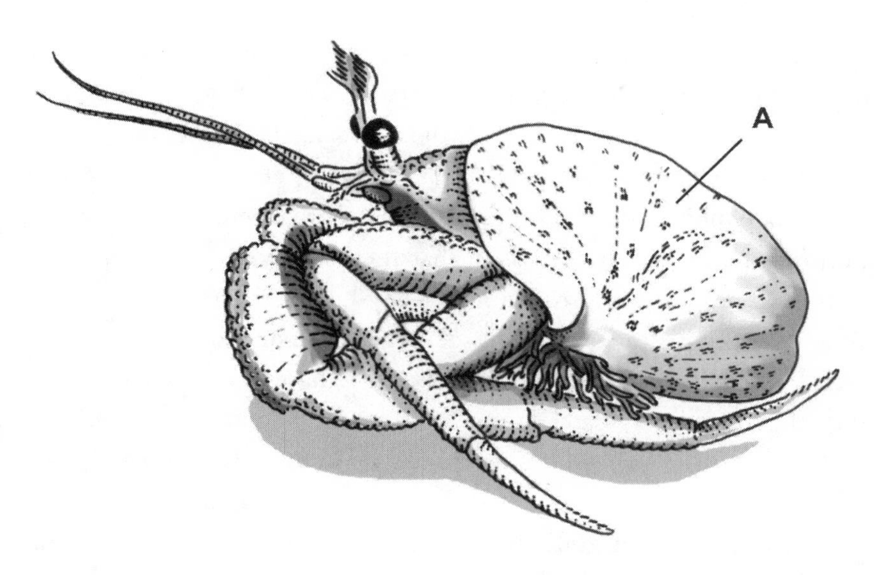

that wood (cellulose) simply could not be digested without the assistance of microbes that live in the gut, as revealed in a review: "the animals that first evolved to feed on cellulose never developed the genes to degrade cellulose, but rather adapted to provide better conditions [in the gut] for microbial [digestion of wood]."[48]

Beware the wording here: Did the animals feeding on cellulose adapt to provide better conditions for microbes that assisted in cellulose digestion? We don't have any evidence that they did. Rather, in the many variations of termite guts that occurred simply because of how replication and variation work, at some time guts that *just happened* to provide a good habitat for cellulose digestion–aiding microbes appeared, and because they were beneficial, the genes for those guts proliferated such that today all termites have them. Over time, then, the termite digestive tract (gut) turned out to be a favorable selective environment for a certain kind of microbe that helps in digesting cellulose, and now the two life-forms—termite and cellulose digester—not only co-exist but co-evolve.[49]

The extreme end of the spectrum of understanding species interactions is presented in the concept of the *hologenome*, which "considers the *holobiont* (the animal or plant with all its associated microorganisms)" as a unit of selection in evolution.[50] In this view, all plants and animals establish symbiotic relationships with microorganisms, and these relationships become engrained such that they *have* to persist if the interacting species are going to survive. They become inseparable, so they simply cannot be understood in isolation. For example, laboratory mice raised in sterile, germ-free conditions tend to have poorer health, slower food digestion, and greater susceptibility to infection than mice allowed to develop with microbial species that all aid in food digestion and so on.

We don't know much about this kind of interaction. As biologist Ernst Mayr wrote in 2002, many species evolve in relationships with others "due to the establishment of consortia between two organisms with extremely different genomes. Ecologists have barely begun to describe these interactions."[51]

Going even further, and combining the revelations of horizontal gene transfer with the discoveries of genomics and co-evolution—and a thoroughgoing conviction that co-evolution has simply been largely ignored in biology—microbiologist Lynn Margulis proposes an entirely new theory about the origin of species. Speciation, Margulis argues, doesn't normally happen by the reproductive isolation and other factors we've seen in this book; rather, for Margulis, it is the acquisition of large sets of DNA by one species, from another, that "drives the bus of evolution."[52] Her recent book, *Acquiring Genomes* (co-authored with Dorion Sagan) is a fascinating and thrilling review of her theory, and I recommend it to anyone interested in an alternative view of evolution that, at the least, describes yet *another* way that evolution occurs. But remember, even Margulis is not ditching the principles of replication, variation, and selection. Those are facts even Margulis accepts. Such facts are not subject to dismissal, and neither is their consequence: the evolution of new life-forms from old over time.

These many fascinating examples give us an intimation of the staggering, almost overwhelming complexity of the world of living things. Imagine, now, trying to understand any life-form: you need to know about its DNA, its development through the life course, its parasites and symbionts, its reaction norm, and its range of variations on the basic anatomical theme; you have to know about its selective environment, its place in the ecosystem, and how that ecosystem is changing, and you have to know about its evolutionary history and its chemistry and even if it has picked up genes from other species somewhere along the line.

WHAT NOW?

Considering these insights, what's left of Darwinian evolution? Has it been overturned? Not quite. What remains are the core principles, and that is significant. Even the renegade biologists Lynn Margulis (whom we've just met) and Donald I. Williamson—each of whom has proposed entirely new modes of evolution—still retain the Darwinian core of replication, variation, and selection. Williamson, in the same article in which he proposes a radical addition to evolutionary theory, even writes: "I do not propose larval transfer as a substitute for natural selection. Adults and larvae have evolved gradually by [Darwinian] "descent with modification," but, superimposed on this process, entire genomes have been transferred [from one kind of life to another] by hybridization."[53]

Even with all this new knowledge, and the calls for a "new biology," what we see is a revision of and an increase in our knowledge of evolution and its many ways, rather than an abandonment of its core factual processes. Deep down, on the molecular level, we know there are things happening that we are only just perceiving, but they do not overturn our understanding of Darwinian evolution.

In a 2009 paper, biologist Eugene V. Koonin reviewed the foundations of biology in the "pre-genomic" and "post-genomic" eras. A few of his points are laid out below:

> *Is variation the principal material for natural selection?*
>
> Yes; but we must remember that mutation has many sources, and that randomness may apply less than we have thought.
>
> *Does natural selection generate increasingly complex adaptations through time?*
>
> In many cases, yes, but not necessarily; maladaptation and extinction are common, and genomes do not necessarily become more complex through time due to a number of constraints on variation.
>
> *Does evolution proceed by making small changes?*
>
> Not necessarily; much evolution occurs on a radically short timetable, as in the case of bottlenecks. Rapid speciation may occur after arrival in previously unoccupied habitats.
>
> *Has evolution always proceeded in the same way?*
>
> To an extent, yes; the essential processes of replication, variation, and selection are required; however early evolution of Earth life was substantially different from that [of] today, and we must consider that evolution itself might have been different at that time.
>
> *Is there such a thing as a species?*
>
> Yes and No; there are clearly different kinds of life, as in the case of an elephant and a cactus; but on the other hand these must be recognized

as shades in a spectrum of DNA-based life-forms. And, among the asexually-reproducing species the entire concept of species must be reconsidered because they gain DNA not just from their "kind" (species) but from other "kinds." Because species never stop evolving, and evolution is characterized by change, we must recognize that the lines drawn around or between species can in some ways be considered arbitrary.[54]

WRAPPING IT UP (UNTIL TOMORROW)

Clearly, recent advances in biological understanding show that evolution has multiple modes. Some evolution is essentially Lamarckian, some evolution is essentially Darwinian, and different modes might have prevailed at different times in the history of life. In sum, there is much more to evolution than the simple Darwinian evolution I present in this book. But those are all elaborations on evolution, not dismissals of its core. Darwinism is not dead, neo-Darwinism is not dead, and evolution is not "just a theory" or flat-out "wrong." Rather, we are better understanding the actual complexity of evolution.

THE FUTURE OF EVOLUTIONARY BIOLOGY

The John Templeton Foundation is a deep pocket from which scientists investigating "the big questions" may, if they're lucky to receive an award, draw big research dollars. Recently the foundation doled out over ten million US dollars to help answer Harvard University's "Foundational Questions in Evolutionary Biology," formulated in 2009 in commemoration of the publishing of Darwin's *On the Origin of Species* 150 years earlier. To give you an idea where in the "big picture" this one (of many) research effort is headed, I've listed some of their questions:

- What is the transition from chemical kinetics to evolutionary dynamics, and can the transition from non-evolving to evolving systems be defined precisely and formally?
- Why does evolution (sometimes) lead to increasing complexity?
- What are the differences between genetic and cultural evolution, and how can these differences be formalized?
- Can we build precise models for the evolution of cells, multicellular organisms, animal societies and human language?[55]

With such great questions to answer, and the powerful new tools like comparative genomics, we are in for quite a show!

Considering the incredible things we've seen in this chapter, what is biology to do now? Carl Woese, arguing that much of genomics has been overly focused on describing gene functions, puts it best:

> A heavy price was [in the past few decades] paid for molecular biology's obsession with metaphysical reductionism [focusing on the gene rather than the gene in its environment]. It stripped the organism from its environment; separated it from its history, from the evolutionary flow; and shredded it into parts to the extent that a sense of the whole—the whole cell, the whole multicellular organism, the biosphere—was effectively gone. Darwin saw biology as a "tangled bank," with all its aspects interconnected. Our task now is to resynthesize biology; put the organism back into its environment; connect it again to its evolutionary past; and feel that complex flow that is organism, evolution, and environment united. The time has come for biology to enter the nonlinear world.[56]

Woese cautions, though, that too much genetic research is ultimately driven by a wish to control the genes and that the direction biology takes from here out is significant not only to understanding living things, but to society itself:

> A society that permits biology to become an engineering discipline, that allows science to slip into the role of changing the living world without trying to understand it, is a danger to itself. Modern society knows that it desperately needs to learn how to live in harmony with the biosphere. Today more than ever we are in need of a science of biology that helps us to do this, shows the way. An engineering biology might still show how to get there; it just doesn't know where "there" is.[57]

CHAPTER 9

THE GRAND ILLUSION

The mind, that Ocean where each kind,
Does straight its own resemblance find,
Yet it creates, transcending these,
Far other Worlds, and other Seas.

—Andrew Marvell, "The Garden"[1]

Despite the essential simplicity of evolution, it remains widely misunderstood and rejected. Obviously a lot of that rejection results from the concept of human centrality in the universe, but I think there's another important reason for it. It has a lot to do with the very way our minds work; with being human itself. That's because the essence of humanness is in the making of things. I believe this proaction—quite unique in the world of living things—has conditioned the human mind and human cultures to believe that complex things (like plants and animals) can only be the result of proactive *making*. To see how, we need to have a look at how the human mind works.

BEING HUMAN

Anthropology (the scientific study of humanity) has shown that there are at least two meanings to "humanness." *Anatomical modernity* means possessing a skeleton that's indistinguishable from the skeletons of modern humans, and we first see this between two hundred thousand and one hundred thousand years ago, in Africa. The other meaning of humanness is *behavioral modernity*, referring to behaviors essentially indistinguishable from those of modern humans; largely, this means the use of complex symbolism and somewhat modern language. Though these can be tough to track archaeologically, there's decent consensus that this is also first seen in Africa, in symbolic artifacts dated to almost eighty thousand years ago.[2] It is behavioral modernity that I'm concerned with here, and what it means for the mind, for how we think, and why the way our minds work can make evolution difficult to accept.

PROACTION, REACTION, AND THE INVENTION OF INVENTION

Behavioral modernity is rooted in planning and creation. While some nonhuman animals make and use tools, humans are entirely dependent on making things—such as a native Arctic seal harpoon or an antibiotic pill—to survive.

This becomes clear when we consider that evolution doesn't look forward; nonhuman offspring of any parent generation are born into environments with bodies more or less identical to those of their parents and therefore more or less suited to the environment in which their parents flourished. If the current environment is different from the parents' environment, the offspring can't do anything about it. Certainly they don't change their bodies to adapt to the new environmental conditions, because they don't know that evolution is happening in the first place, and because there's just no way to rapidly tailor the physical body to fit new environments. And while we saw that bodies can change in reaction to environmental cues, again, those reactions are not consciously controlled; they're *re*actions, not *pro*actions.

But human evolution is very different. Humanity's trick—and it's a diabolically good one—is the ability to quickly adjust to any selective pressure by inventing adaptations. In this way, behaviorally modern humans are proactive, not just reactive. Inventions can be artifacts, like a pair of warm boots; or complex behaviors, such as a dance that symbolically reminds people of the details of how to hunt a particular animal. Whatever the invention, the point is that people thought it up; they perceived a problem and then designed a solution to fit that problem. And we don't just do it for fun; we live or die by our ability to buffer our frail bodies from an ever-changing array of selective pressures. This has allowed human behavior to become—in paleontologist David Pilbeam's words—largely "decoupled from our anatomy."[3]

I think this is one of the most important concepts in all of human evolution, because, ultimately, humanity's most potent adaptation has been the *invention of invention*. Inventing adaptations or variations that increase your fitness in the terms we've used in this book means that you don't have to wait for biological variations to come up before adapting. This basically allows humanity to adapt instantly rather than on a biological timescale. While this proactive adaptation began over two million years ago, with the use of stone tools to do what our frail bodies could not do, today we are entirely dependent on our artifacts—our made objects—to survive. Australian aborigines have a dozen kinds of stone and wood tools necessary for survival, and of course the world's great cities would quickly collapse without the thousands of inventions that keep them supplied with water, food, and so on.

Our making and use of things is deeply rooted in our consciousness, and the enculturation begins early: from the day we learn that a peanut-butter-and-jelly sandwich doesn't just spontaneously assemble itself—that it has to be assembled with intent (preferably by someone else)—it just seems obvious that all the other things we see in the world (or at least those at least as complex as a peanut-butter-and-jelly sandwich) were similarly assembled *with intent*. An acorn, for example, or a sturgeon or an owl or a virus; each is such a wonder of design (you try to make one!) that they must have been made by intent. And so humanity has thought for a long, long time, because—for as long as we can remember—humanity has been intentionally making things for specific purposes.

The preconception that everything must have been made by intent has a second source, in addition to our reliance on made tools. In addition to a reliance on inventions to survive, the complex communication system that we call language is another important criterion of humanness. And when we compare human and nonhuman communications, we see that complex human language is uniquely proactive and that use of language is a deeply creative process.

TEFLON, VELCRO, AND THE MODERN HUMAN MIND

Human language is a complex system of animal communication. While other animals certainly communicate, humans use many complex rules to exchange large amounts of information with great subtlety, at high speed, and with relatively few errors. Figure 9-1 shows gestural (bodily) communication in the nonhuman and human worlds. We see that dogs (A) indicate aggression or submission by very clear "opposite" postures, and elephants (B) communicate various states of arousal by flapping their ears and touching the trunk to the forehead. Humans also use body language in very subtle ways; (C) is derived from a nineteenth-century painter's manual, depicting different ways of holding objects as well as different

Fig. 9-1. Bodily Communication in Humans and Other Animals

body language among different social classes. While humans communicate in ways similar to other life-forms, what is most distinctive about human language is the complexity of its rules—the grammar—and the nature of its symbols, which actively promote invention and creation.

Nonhuman primate communication uses the simplest kind of symbolism, as when a monkey gives an "aerial predator" screech or "ground predator" hoot, eliciting the proper defensive behaviors from its companions. These vocalizations are symbols because the sounds are arbitrary and they "mean" something else; one vocalization means "ground predator"; another means "air predator." What is most interesting about these communications is that they're very simple; they use very simple symbols, which I call "Teflon" because nothing sticks to them—the aerial predator screech can only ever mean aerial predator. The symbol to symbolized ratio is 1:1.[4]

In the most profound contrast, humans "stick" concepts together, making more complex messages; this is what I call "Velcro" symbolism, because one symbol sticks to the next. For example, we might say, "Watch out for that guy, he's a real snake!" With human language, the symbol to symbolized ratio is 1:*n* because any word can be used to mean anything else we choose. The sound used to indicate "snake" can now be applied to the characteristics of another person. Somewhere in the evolution of our minds, our lineage broke some kind of barrier such that symbols were no longer concretely attached to one another; anything could mean anything.[5]

What's the advantage of such complex language, of "Velcro" symbolism? At the very least, it allows a better "fit" between individuals (or groups) and their world. Simple, slavish reactions to screeches that mean "Climb fast!" (ground predator) or "Drop fast!" (aerial predator) may be wasteful and unnecessary. But more specific reactions tailored to subtler messages mean better reactions. Also, the increased complexity of the human language system allowed more (and more detailed) information to be communicated rapidly.

In the behaviorally modern human mind, then, we find the capacity for infinite communication novel concepts, due to 1:*n* symbol to symbolized ratio and "Velcro" idea association. The combination has staggering power and has had staggering results; a once-obscure bipedal primate now effectively runs the planet.

It's fascinating that (counterintuitively, for me, anyway) subtlety in our communications is promoted not by more *rigidly* defining what certain sounds (symbols) mean, but by promoting innovation (invention) of subtler, more significant messages by sticking one symbol to another! For example, humans can assemble a metaphor, as we saw before, saying, "That guy is a real snake." This even *mixes* information from the nonhuman (snake) and social (guy) domains of thought.

This capacity for "Velcro" symbolism is what really distinguishes humans from other animals, and from its origin—at least one hundred thousand years ago—humanity has been making things (including sentences) with conscious intent. Even receiving a message requires deciphering exactly what the other person really meant as opposed to (perhaps) what they actually said, so even the act of interpretation, shaped by the knowledge of intended creation by the other party, is itself purposive creation close to the very heart of humanness.[6]

So, from the beginning of what we can call behavioral modernity, we humans have been surviving by intent and sticking one idea to another, sometimes proactively to invent solutions, and we have come to value proaction, planning, and making.

CREATIVITY AND THE UNDERSTANDING OF EVOLUTION

Creating, making, inventing, building, imagining—everything from stone tools to poems—are proactive human acts. Our mythologies are full of creation. We venerate exceptional creators. There is no end to human creation, so long as there are minds capable of "sticking" ideas and words together in new combinations. Creativity is at the heart of humanness.

This is why it can be so hard to understand that evolution is not, also, a result of proaction. But when we understand that nonhuman evolution isn't actually a thing but rather the completely unintended consequence of three independent, factual, and observable processes (replication of life-forms, variation in offspring, and selection among offspring), we gain a far richer understanding of how the natural world works and how all these wonderful slime molds, birches, Venus flytraps, and jellyfish, and every other living thing, actually came to be.

NOTES

FOREWORD

1. Thomas Bulfinch, *Bulfinch's Greek and Roman Mythology: The Age of Fable* (Mineola, NY: Dover, 2000).

2. The quotation is from the 1873 novel *Middlemarch: A Study of Provincial Life* by George Eliot (pen name of Mary Anne Evans), 1819–1880. You can read more—much more—about the influence of Darwinism on British literature in G. Beer, *Darwin's Plots: Evolutionary Narrative in Darwin, George Eliot, and Nineteenth-Century Fiction*, 3rd ed. (Cambridge: Cambridge University Press, 2009).

INTRODUCTION

1. I don't mean to beat up on Matt Ridley's excellent science writing, but I do think this is a significant miswording: see M. Ridley, *Genome: The Autobiography of a Species in 23 Chapters* (New York: Harper Perennial, 2006), p. 33.

CHAPTER 1: *NULLIUS IN VERBA*

1. See A. Kehoe, "Modern Antievolutionism: The Scientific Creationists," p. 181 in *What Darwin Began: Modern Darwinian and Non-Darwinian*

Perspectives on Evolution, ed. L. R. Godfrey (Boston: Allyn & Bacon, 1985), pp. 156–85.

2. For over 140 years, the British journal *Nature* has carried articles on discoveries in science. Originally (in 1869) intended to "place before the general public the grand results of scientific work and scientific discovery; and to urge the claims of science to move to a more general recognition in education and in daily life," today the journal mostly carries highly technical research articles that can be understood only by scientists themselves; this is because of the proliferation of science, not as a way to keep knowledge from the public. *Nature* (and the other science journals I refer to throughout this book) is a peer-reviewed publication, meaning that articles submitted by scientists are reviewed by others in their field (peers) for logical consistency, adequate methods, and reasonable inferences before they are approved for publication. An early proponent of science's self-correcting mechanism of peer review was British mathematician Charles Babbage (1791–1871); see R. Holmes, *The Age of Wonder* (New York: Vintage Books, 2009) p. 439. Peer review, although imperfect, is used to maintain high standards in science. For the editorial quotation, see the editorial "Spread the Word" in *Nature* 451:108, at http://www.nature.com/nature/journal/v451/n7175/full/451108b.html (accessed March 10, 2011).

3. N. Eldredge, "Evolutionary Tempos and Modes: A Palaeontological Perspective," in *What Darwin Began: Modern Darwinian and Non-Darwinian Perspectives on Evolution*, ed. Laurie Rohde Godfrey (Boston: Allyn & Bacon, 1985), pp. 113–17.

4. A recent article in the *Guardian* newspaper is titled "Why Everything You've Been Told about Evolution Is Wrong." The story doesn't show that everything you've been told about evolution is wrong; the story says that what you've been told "isn't the whole story." But that is normal; since science is constantly updating what it knows, there never is a "whole story." See Oliver Burkeman, "Why Everything You've Been Told about Evolution Is Wrong," *Guardian*, March 19, 2010, http://www.guardian.co.uk/science/2010/mar/19/evolution-darwin-natural-selection-genes-wrong (accessed March 10, 2011).

5. The quotation is from a video documentary (*Solving the Puzzle*) featuring Carl Woese (in part 2 of 10) on the microbiology.org website. See http://www.microbeworld.org/ and click on "Video" (accessed March 10, 2011).

6. The Royal Society, the "UK academy of science promoting the natural and applied sciences, as a learned society, and as a funding agency" (http://royalsociety.org/about-us/) was founded in London in 1660.

7. I am not saying that there has been no debate over evolution in Britain, or that no American scientists accepted evolution as fact very early on. But there is a tangible, cultural difference in the "touchiness" of evolution as a topic, even in general conversation, between Britain (and Europe in general) and the United States.

8. T. H. Huxley, *Darwiniana: Essays by Thomas Henry Huxley*, vol. 2 (London: Macmillan, 1863), p. 467.

9. See R. L. Numbers, *Darwinism Comes to America* (Cambridge: Harvard University Press, 1998), p. 3.

10. To avoid the "black hole" topic of religiously based critiques of evolution (or of any branch of science), here I refer you to two books; Eugenie Scott's *Evolution and Creationism* (University of California Press, 2008) is a thorough (nearly four-hundred-page) treatment of the issue. My own book, coauthored with Charles Sullivan (*The Top Ten Myths about Evolution* [Amherst, NY: Prometheus Books, 2006]) more rapidly dismantles some common religious critiques of evolution.

11. See *Science, Evolution, and Creationism*, National Academy of Sciences, Washington, DC, 2008, http://www.nap.edu/catalog/11876.html (accessed March 10, 2011).

12. Ibid.

13. Adam seems happy to comply: "To whom thus Adam, cleared of doubt, replied / How fully hast thou satisfied me, pure Intelligence of Heaven, Angel serene! / And, freed from intricacies, taught to live / The easiest way; nor with perplexing thoughts . . ." J. Milton, *The Annotated Milton: Complete English Poems*, ed. B. Raffel (New York: Bantam, 1999), p. 374.

14. See http://www.aaas.org/news/press_room/evolution/qanda.shtml. The American Association for the Advancement of Science was founded in 1848 as "an international non-profit organization dedicated to advancing science around the world by serving as an educator, leader, spokesperson and professional association," http://www.aaas.org/aboutaaas/ (accessed September 10, 2010).

15. See *Science, Evolution, and Creationism.*

16. See "DNA Agrees with All the Other Science: Darwin Was Right," *Discover*, March 2009, http://discovermagazine.com/2009/mar/19-dna-agrees-with-all-the-other-science-darwin-was-right (accessed March 10, 2011).

CHAPTER 2: THE FACT OF REPLICATION

1. *Lucretius: On the Nature of the Universe*, trans. and ed. R. E. Latham (London: Penguin Books, 1982), p. 133.

2. K. Ruiz-Mirazo, J. Pereto, and A. Moreno, "A Universal Definition of Life: Autonomy and Open-Ended Evolution," *Origins of Life and Evolution of the Biosphere* 34 (2004): 323–46, and E. C. Minkoff, *Evolutionary Biology* (Reading, MA: Addison-Wesley, 1983).

3. B. Guttman et al. put it this way: "Information is what you get when you specify one thing out of a range of possibilities." B. Guttman et al., *Genetics: A Beginner's Guide* (Oxford: Oneworld, 1982), p. 47.

4. I. Zachar and E. Szathmary, "A New Replicator: A Theoretical Framework for Analysing Replication," *BioMed Central Biology* 8, no. 21 (2010): 1–26, http://www.biomedcentral.com/1741-7007/8/21 (accessed August 23, 2010, especially fig. 13).

5. For more on replicators, see R. Dawkins, *The Extended Phenotype: The Gene as the Unit of Selection* (San Francisco: W. H. Freeman, 1982), and R. Dawkins, "Universal Darwinism, " in *Evolution: From Molecules to Men*, ed. D. S. Bendall, 403–425 (Cambridge: Cambridge University Press, 1983).

6. The other self-replicators are related to DNA or are synthetic self-replicating molecules built in laboratories; for one example, see "Artificial Molecule Evolves in Lab," doi:10.1126/science.1167856 (accessed August 10, 2010).

7. It is unclear exactly when and where citrus fruits were first domesticated, but it was somewhere in East Asia on the order of six thousand years ago. See D. Brothwell and P. Brothwell, *Food in Antiquity: A Survey of the Diet of Early Peoples* (Baltimore: Johns Hopkins University Press, 1969), p. 51. In coming years, new molecular techniques will undoubtedly "nail down" the time and place of early domestication of many species.

8. Beware the of the false "common-sense" impression, though, that everyone worldwide "switched" from foraging (hunting and gathering) to farming ten thousand years ago. Many populations, in fact, continued to forage, and there are still some human societies that hunt (for example, Alaskan and Canadian Inuit people), for example, or raise domesticated animals (for example, Maasai people of East Africa) for their daily subsistence. I am currently writing a book about the prevalence of such common myths about the ancient world.

9. Some say that selective breeding—picking which animals or plants to propagate and (basically) eating the others—is the earliest form of genetic engineering. Technically this is the case, but what modern genetic engineering does to domesticated crops and animals is very different; rather than work with naturally occurring plants and animals, modern genetic engineering actually manipulates the DNA of these organisms. The pros and cons of this practice should become evident as this book unfolds.

10. The word *heredity* derives from the Latin for "the condition of being an heir." Since 1903, the *Journal of Heredity* has carried scientific articles on "organismal genetics: gene action, regulation, and transmission in both plant and animal species, including the genetic aspects of botany, cytogenetics and evolution, zoology, and molecular and developmental biology," http://jhered.oxfordjournals.org (accessed July 19, 2010). Note that before 1905, heredity was less a scientific concern than one of farmers, who had been breeding plants and animals for millennia.

11. For a masterful account of the Middle Ages, see B. Tuchman, *A Distant Mirror: The Calamitous Fourteenth Century* (New York: Ballantine, 1987). For an account of how Irish monks preserved much ancient knowledge, see T. Cahill, *How the Irish Saved Civilization*, Hinges of History Series, vol. 1 (New York: Anchor Books, 1996).

12. See C. Pinto-Correia, *The Ovary and Eve: Egg and Sperm and Preformation* (Chicago: University of Chicago Press, 1998), specifically. For a more general account, see I. Maclean, *The Renaissance Notion of Women: A Study in the Fortunes of Scholasticism and Medical Science in European Intellectual Life* (New York: Cambridge University Press, 1980).

13. Note that neither Hartsoeker nor his contemporary, "Father of the Microscope" Antony van Leeuwenhoek (1632–1723), actually claimed to have seen these little people; Pinto-Correia shows that the common association of the term *homunculus* ("little person")—and the oversimplification associated with it—is not attributable to Hartsoeker or van Leeuwenhoek but was associated with their names by later authors in an attempt to belittle the concept. See Pinto-Correia, *The Ovary and Eve*.

14. In an 1867 letter to Thomas Huxley, Darwin struggled to find a word to convey the fact that cells could "throw off an atom of its contents," suggesting the word *gemmule*, among others. See G. L. Geison, "Darwin and Heredity: The Evolution of His Hypothesis of Pangenesis," *Journal of the History of Medicine and Allied Sciences* 24, no. 4 (1969): 375–411.

15. F. Galton, "Experiments in Pangenesis, by Breeding from Rabbits of a Pure Variety, Into Whose Circulation Blood Taken from Other Varieties Had Previously Been Transfused," *Proceedings of the Royal Society* 19 (1871): 393–410.

16. J. C. Howard, "Why Didn't Darwin Discover Mendel's Laws?" *Journal of Biology* 8, no. 15 (2009), doi:10.1186/jbiol123 (accessed March 10, 2011), italics in the original.

17. A nice summary of Flemming's work, along with some of his careful illustrations, is available in N. Paweletz, "Walther Flemming: Pioneer of Mitosis Research," *Nature Reviews Molecular Cell Biology* 2 (2001): 72–75.

18. J. D. Watson with A. Berry, *DNA: The Secret of Life* (New York: Alfred A. Knopf, 2003), p. 88.

19. It may appear that sometimes a "blend" of two characteristics appears in the offspring, but that is the appearance of a variation that already exists (rather than is a blend) but just doesn't come up too often; for this, it's called a *recessive* characteristic.

20. One-time vice presidential candidate Sarah Palin recently mocked scientific "projects that have little or nothing to do with the public good. Things like fruit fly research in Paris, France. I kid you not." (transcript available at http://www.npr.org/templates/story/story.php?storyId=113870272). Considering that Palin has a child with Down syndrome, and that science based on fruit-fly research is used in the understanding of this and other disorders, Palin's comment is tragically ignorant.

21. Award Ceremony Speech: Presentation Speech by F. Henschen, member of the Staff of Professors of the Royal Caroline Institute, on December 10, 1933, Nobel Prize in Physiology or Medicine 1933, http://nobelprize.org/nobel_prizes/medicine/laureates/1933/press.html (accessed March 10, 2011).

22. F. Crick, *What Mad Pursuit* (New York: Basic Books, 1990), p. 66.

23. You can see Franklin's "Photo 51" online at http://www.pbs.org/wgbh/nova/photo51/pict-01.html. Franklin, who died in 1958, independently made a substantial contribution to understanding the structure of DNA and was credited with this in the journal *Nature*, but her full contribution to understanding DNA has always been eclipsed by attention paid to Watson and Crick; see R. Olby, *The Path to the Double Helix* (London: Dover, 1994).

24. J. D. Watson, *The Double Helix* (New York: Atheneum, 1968), p. 167.

25. See p. 95 of F. H. C. Crick and J. D. Watson, "The Complementary Structure of Deoxyribonucleic Acid," *Proceedings of the Royal Society of London* A 223, no. 1152 (1954): 80–96.

26. Watson, *The Double Helix*, pp. 72–76.

27. M. Ridley, *Genome: The Autobiography of a Species in 23 Chapters* (New York: Harper Perennial, 1999), p. 7.

28. S. P. Doolan and D. W. McDonald, "Breeding and Juvenile Survival among Slender-Tailed Meerkats (*Suricata suricatta*) in the South-Western Kalahari: Ecological and Social Influences," *Journal of Zoology* 242 (1997): 309–27.

29. A wonderful review of the world of chemical communication in animals is found in T. D. Wyatt, *Pheromones and Animal Behavior: Communication by Smell and Taste* (Cambridge: Cambridge University Press, 2008).

30. E. O. Wilson, *Sociobiology: The New Synthesis* (Cambridge: Belknap/Harvard University Press, 1975), p. 179.

31. P. G. D. Feulner et al., "Electrifying Love: Electric Fish Use Species-Specific Discharge for Mate Recognition," *Biology Letters* 5, no. 2 (2009): 225–28.

32. R. T. Hanlon, "Mating Systems and Sexual Selection in the Squid *Loligo*: How Might Commercial Fishing on Spawning Squids Affect Them?" *CalCOFL Report* 39 (1998): 92–100.

33. S. R. Telford and P. I. Webb, "The Energetic Cost of Copulation in a Polygynandrous Millipede," *Journal of Experimental Biology* 201 (1998): 1847–49.

34. R. C. Babcock, C. N. Mundy, and D. Whitehead, "Sperm Diffusion Models and In Situ Confirmation of Long-Distance Fertilization in the Free-Spawning Asteroid *Acanthaster planci*," *Biological Bulletin* 186, no. 1 (1994): 17–28.

CHAPTER 3: THE FACT OF VARIATION

1. R. E. Latham, trans. and ed., *Lucretius: On the Nature of the Universe* (London: Penguin Books, 1982).

2. The illusion was that of the "Great Chain of Being," which hierarchically organized all life-forms according to their proximity to the gods (who were at the top of the chain). For more on the idea and how it structured Western thought, see C. M. Smith and C. Sullivan, *The Top Ten Myths about*

Evolution (Amherst, NY: Prometheus Books, 2006), especially chap. 3, "The Ladder of Progress," or—for an in-depth treatment—see A. O. Lovejoy, *The Great Chain of Being: A Study of the History of an Idea* (Cambridge, MA: Harvard University Press, 1936).

3. The quotation is from E. H. Grell, K. B. Jacobson, and J. B. Murphy, "Alterations of Genetic Material for Analysis of Alcohol Dehydrogenase Isozymes of Drosophila melanogaster," *Annals of the New York Academy of Sciences* 151 (1968): 441–45, as quoted in M. Ashburner, "Speculations on the Subject of Alcohol Dehydrogenase and Its Properties in Drosophila and Other Flies," *BioEssays* 20 (1998): 949–54.

4. Y. Malherbe et al., "ADH Enzyme Activity and Adh Gene Expression in *Drosophila melanogaster* Lines Differentially Selected for Increased Alcohol Tolerance," *Journal of Evolutionary Biology* 18 (2005): 811–19.

5. R. Dawkins, *The Extended Phenotype: The Gene as the Unit of Selection* (Oxford and San Francisco: W. H. Freeman, 1982), p. 114.

6. The idea of the *inheritance of acquired characteristics* is normally identified with French naturalist Jean-Baptiste Lamarck (1744–1829), who proposed a different variety of the evolution of species through time than Darwin; Darwin turned out to be correct, although, as we'll see in chapter 8, a variety of "quasi-Lamarckian" evolution does occur in some living species.

7. L.-A. Gershwin, "Clonal and Population Variation in Jellyfish Symmetry," *Journal of the Marine Biological Association of the United Kingdom* 79 (1999): 993–1000. My figure is an adaptation from work in that report.

8. E. C. Minkoff, *Evolutionary Biology* (Reading, MA: Addison-Wesley, 1983), fig. 6.5.

9. Ibid., fig. 13.7.

10. My figure is adapted from G. Edelman, *Bright Air, Brilliant Fire* (New York: Basic Books, 1993), figs. 3–5 (locusts); K. C. Dodd, *American Box Turtles: A Natural History* (Norman: University of Oklahoma Press, 2002), figs. 1–4 (turtles); and R. van Zyll de Johng, H. Reynolds, and W. Olson, "Phenotypic Variation in Remnant Populations of North American Bison," *Journal of Mammalogy* 76 no. 2 (1995): 391–405, figs. 1 and 2 (bison).

11. M. D. Abrams, "Genotypic and Phenotypic Variation as Stress Adaptations in Temperate Tree Species: A Review of Several Case Studies," *Tree Physiology* 14, no. 7 (1994): 833–42.

12. M. B. Sokolowski, "Genes for Normal Behavioral Variation: Recent Clues from Flies and Worms," *Neuron* 21 (1998): 463–66. For a long time, the *correlation* between letting go of an object, and its fall to earth, was well documented, but the *causal explanation* (the law of gravitation indicating that a larger mass—the earth—will attract a lesser mass—the dropped object) was not well understood. In many ways, with such studies, we are today, biologically, in the same position. That should remind us that it may be possible to identify the causal explanation. As ever, in science, who knows what tomorrow's research journals will clarify.

13. In a short review of several cases of behavioral variation being correlated with gene-level variation, Sokolowski reminds us that some of the ambiguity in current research results because "we just don't know enough about how to identify genes important to behavioral variation." See Sokolowski, "Genes for Normal Behavioral Variation," p. 463. That doesn't mean we *can't* know, only that we don't at the moment. You can bet that plenty of research is under way, at this very moment and across the globe, to improve our understanding.

14. J. K. Webb, G. P. Brown, and R. Shine, "Body Size, Locomotor Speed, and Antipredator Behaviour in a Tropical Snake (*Tropidonophis mairii, Colubridae*): The Influence of Incubation Environments and Genetic Factors," *Functional Ecology* 15 (2001): 561–68.

15. D. Nettle, "The Evolution of Personality Variation in Human and Other Animals," *American Psychologist* 61, no. 6 (2006): 624–25. This same study explores the significance of such studies for human evolution, especially pp. 622–31.

16. Naturalist and author Barry Lopez has emphasized that attaining wisdom about the natural world—a full knowledge and understanding of a given ecosystem—takes significant time and effort, often more than our civilization, largely focused on the short-term, finds acceptable. Evolutionary biol-

ogist John Endler, among many others, has also called for more comprehensive, long-term field studies for exactly this reason (J. A. Endler, *Natural Selection in the Wild* [Princeton, NJ: Princeton University Press, 1986], generally, and p. 247 specifically). I'm happy to report that many modern studies recognize exactly this point and are transgenerational; for example, the many studies of chimpanzee and gorilla behavior that have been under way for over forty years. See B. Lopez, "The Naturalist," *Orion* magazine, 2001, http://www.orion magazine.org/index.php/articles/article/91/ (accessed March 10, 2011).

17. We'll revisit the question of the ubiquity of variation later, in chapter 8. For the moment, consider that many sources of variation have been found in microbial life in the past few decades, prompting even a 1993 paper titled "How Clonal Are Bacteria?" See J. M. Smith et al., "How Clonal Are Bacteria?" *Proceedings of the National Academy of Sciences of the United States of America* 90 (1993): 4384–88.

18. See p. 577 of S. B. Carroll, "Endless Forms: The Evolution of Gene Regulation and Morphological Diversity," *Cell* 101 (2000): 577–80.

19. Gould and Lewontin (see note below) didn't pick a German term by accident; German biology had a long history of considering this kind of subtlety in evolution, subtlety that was distinctly lacking in the more mechanistic, atomized concepts of life in American and British biology. This Germanic "organic" conception of life and the universe had many effects even outside biology, per se: see, for example, its effects on astronomy in R. Holmes, *The Age of Wonder: How the Romantic Generation Discovered the Beauty and Terror of Science* (New York: Pantheon Books, 2009).

20. See S. J. Gould and R. C. Lewontin, "The Spandrels of San Marco and the Panglossian Paradigm: A Critique of the Adaptationist Programme," *Proceedings of the Royal Society of London* B 205 (1979): 581–98. Note that Gould and Lewontin were writing at least a decade before details of the genome were even examinable, because of the technological constraints of their time. Even twelve years later, Gould wrote that the molecular basis of different bauplan remained "only dimly understood." (S. J. Gould, "The Disparity of the Burgess Shale Arthropod Fauna and the Limits of Cladistic Analysis," *Paleobiology* 17

[1991]: 411–23), whereas today that molecular basis—as we see throughout this book—is available, and an object of intense study (see chapter 8).

21. P. Chai and D. Millard, "Flight and Size Constraints: Hovering Performance of Large Hummingbirds under Maximal Loading," *Journal of Experimental Biology* 200 (1997): 2757–63, http://jeb.biologists.org/cgi/content/abstract/200/21/2757. It is fascinating that hummingbirds produce lift not only on the downstroke, but also on the wing's upstroke, which provides about a quarter of the total lift. See D. R. Warrick, B. W. Tobalske, and D. R. Powers, "Aerodynamics of the Hovering Hummingbird," *Nature* 435 (2005): 1094–97. See also D. J. Wells, "Muscle Performance in Hummingbirds," *Journal of Experimental Biology* 178 (1993): 39–57.

22. F. Santos et al., "Dynamic Reprogramming of DNA Methylation in the Early Mouse Embryo," *Developmental Biology* 241, no. 1 (2002): 172–82.

23. J. Roux and M. Robinson-Rechavi, "Developmental Constraints on Vertebrate Genome Evolution," *PLoS Genet* 4, no. 12:e1000311, doi:10.1371/journal.pgen.1000311 (accessed March 10, 2011).

24. J. D. Hatle, D. W. Borst, and S. A. Juliano, "Plasticity and Canalization in the Control of Reproduction in the Lubber Grasshopper," *Integrative Comparative Biology* 43 (2003): 635–45.

25. See E. C. Friedberg, G. C. Walker, and W. Seide, *DNA Repair and Mutagenesis* (Washington, DC: ASM, 1995) and several articles in the December 23, 1994, issue of *Science*.

26. Another review of limits to variation—which names low population size, low mutation rates, and "fixation," also known as "canalization" (as in a narrow canal where water flows only in one very specific direction), of the genotype and the bodies built by it—can be found in M. W. Blows and A. A. Hoffman, "A Reassessment of Genetic Limits to Evolutionary Change," *Ecology* 86, no. 6 (2005): 1371–84. Yet another is found in S. J. Arnold, "Constraints on Phenotypic Evolution," *American Naturalist* 140, suppl. 1 (1992): S85–S107. A case of "canalization" is found in the wing-blood-supply pattern of the fruit fly, in which, for over fifty million years, it appears that the basic wing-blood-supply pattern of five long horizontal vessels (that reach from the

origin of the wing, near the body, to the tip of the wing) supply several basically perpendicular blood vessels as observed in 2,774 studied individuals of 23 different species, from rain forest dwellers to desert dwellers and so on. See T. F. Hansen and D. Houle, "Evolvability, Stabilizing Selection, and the Problem of Stasis," in *Phenotypic Integration: Studying Ecology and the Evolution of Complex Phenotypes*, ed. M. Pigliucci and K. Preston (Oxford: Oxford University Press, 2004), pp. 130–50.

27. R. Frankham, "Do Island Populations Have Less Genetic Variation Than Mainland Populations?" *Heredity* 78 (1997): 311–27.

28. K. E. Schwaegerle and B. A. Schaal, "Genetic Variability and the Founder Effect in the Pitcher Plant *Sarracenia purpurea* L.," *Evolution* 33, no. 4 (1979): 1210–18. Bottlenecks are well recognized scientifically, but note that things are not always so simple: "Results of recent empirical studies suggest that while genetic variation may decrease with reduced remnant population size, not all fragmentation events lead to genetic losses and different types of genetic variation . . . may respond differently." See p. 413 of A. Young, T. Boyle, and T. Brown, "The Population Genetic Consequences of Habitat Fragmentation for Plants," *TREE* 10, no. 11 (1998): 413–18.

29. M. Menotti-Raymond and S. J. O'Brien, "Dating the Genetic Bottleneck of the African Cheetah," *Proceedings of the National Academy of Sciences* 90, no. 8 (1993): 3172–76. See also S. J. O'Brien et al., "Genetic Basis for Species Vulnerability in the Cheetah," *Science* 227 no. 4693 (1985): 1428–34.

CHAPTER 4: THE FACT OF SELECTION

1. *Lucretius: On the Nature of the Universe*, trans. and ed. R. E. Latham (London: Penguin Books, 1982).

2. For the moment I am ignoring—at my peril, I am sure—the recent building of synthetic DNA; see "Craig Venter Creates Synthetic Life Form," http://www.guardian.co.uk/science/2010/may/20/craig-venter-synthetic-life-form (accessed September 5, 2010).

3. In this book I am using practical terms that communicate the essence of the issue at hand; any point can be investigated in apparently infinite detail. Regarding natural selection, I have been guided by several good definitions that are too technical to include in the text but are interesting here. John Endler, in *Natural Selection in the Wild* (Princeton, NJ: Princeton University Press, 1986), pp. 4–6, states that natural selection is a process in which heritable characteristics of a population will change over time if those characteristics affect the life-form's fitness [measurable as the "average lifetime contribution to the breeding population"]). In *Natural Selection* (New York: Oxford University Press, 1992), p. 5, George C. Williams states that "[n]atural selection is a system of corrective feedback that favors those individuals that most closely approximate some best available organization of their ecological niche." In *Evolutionary Biology* (Reading, MA: Addison-Wesley, 1983), p. 82, Eli C. Minkoff refers to natural selection as "the differential contribution of heritable variations to the next generation." I could go on, but it is from these three approaches—for all kinds of reasons—that I build my phrasing in this book.

4. N. B. Frazer, "Sea Turtle Conservation and Halfway Technology" *Conservation Biology* 6, no. 2 (2003): 179–84.

5. R. A. Sweitzer and D. H. Van Vuren, *Rooting and Foraging Effects of Wild Pigs on Tree Regeneration and Acorn Survival in California's Oak Woodland Ecosystems*, USDA Forest Service General Technical Report PSW-GTR-184, 2002.

6. T. Larsen, "Polar Bear Denning and Cub Production in Svalbard, Norway," *Journal of Wildlife Management* 49, no. 2 (1985): 320–26.

7. W. C. Webb, W. J. Boarman, and T. Rotenberry, "Common Raven Juvenile Survival in a Human-Augmented Landscape," *Condor* 106 (2004): 517–28.

8. Darwin first used the term "survival of the fittest" in the title to chapter 4 ("Natural Selection; or The Survival of the Fittest") in the sixth edition (considered the definitive) of *On the Origin of Species*, published in 1872, thirteen years after the first edition. He took the phrase from Herbert Spencer (1820–1903), a prominent English intellectual.

9. For the twenty-eight slightly different uses of the term *fitness*, see appendix B in R. E. Michod, *Darwinian Dynamics: Evolutionary Transitions in Fitness and Individuality* (Princeton, NJ: Princeton University Press, 1999), pp. 222–25, which includes the excellent phrasing of T. Dobzhansky, relating fitness to the "ability of the organism to survive and reproduce in an environment," p. 223. Michod's phrase "expected reproductive success" also relates to the probability of passing one's genes on to the next generation, which is the crux of the matter. See Michod, *Darwinian Dynamics*, p. 176. Chap. 10 of Richard Dawkins, *The Extended Phenotype: The Gene as the Unit of Selection* (Oxford and San Francisco: W. H. Freeman, 1982), is all about fitness.

10. Endler, *Natural Selection in the Wild*, pp. 33–51.

11. Dawkins, *The Extended Phenotype*, p. 185.

12. Incredibly, blind humans have learned to use echolocation to help them get around; see "Echolocation Allows Blind Humans to 'See,'" http://dsc.discovery.com/news/2009/07/07/human-echolocation.html (accessed September 14, 2010). For an example of whale echolocation, see P. T. Madsen et al., "Bisonar Performance of Foraging Beaked Whales (*Mesoplodon densirostris*)," *Journal of Experimental Biology* 208 (2005): 181–94.

13. Minkoff, *Evolutionary Biology*, p. 82.

14. E. Mayr, *What Makes Biology Unique? Considerations on the Autonomy of a Scientific Discipline* (Cambridge: Cambridge University Press, 2004), p. 135.

15. C. M. Smith and C. Sullivan, *The Top Ten Myths about Evolution* (Amherst, NY: Prometheus Books, 2006), pp. 13–14.

16. Not only do big raptors prey on modern primates, but they have done so for millions of years. The damage to the primate bones found in this study corroborates a previous study claiming that two- to three-million-year-old early hominin fossils from South Africa showed evidence—in the form of punctures and other marks on the bones—of large raptor predation. See W. S. McGraw, C. Cooke, and S. Shultz, "Primate Remains from African Crowned Eagle (*Stephanoaetus coronatus*) Nests in Ivory Coast's Tai Forest: Implications for Primate Predation and Early Hominid Taphonomy in South Africa," *American*

Journal of Physical Anthropology 131 (2006): 151–65. A good overview of the whole issue of hominins being the prey of other species is found in D. Hart and R. W. Sussman, *Man the Hunted: Primates, Predators, and Human Evolution* (Boulder, CO: Westview Press, 2005).

17. N. E. Pierce and P. S. Mead, "Parasitoids as Selective Agents in the Symbiosis between Lycaenid Butterfly Larvae and Ants," *Science* 211 (1981): 1185–87.

18. C. Portier et al., "Effects of Density and Weather on Survival of Bighorn Sheep Lambs (*Ovis canadensis*)," *Journal of Zoology* 245 (1998): 271–78.

19. See chap. 3, n. 16.

20. Very confusingly, negative selection is also variously referred to as "purifying," "centripetal," or "stabilizing" selection. See p. 85 L. D. Hurst, "Genetics and the Understanding of Selection," *Nature Reviews Genetics* 10 (2009): 83–93.

21. N. Waser and M. V. Price, "Pollinator Choice and Stabilizing Selection for Flower Color in *Delphinium nelsonii*," *Evolution* 35, no. 2 (1983): 376–90.

22. Very confusingly, positive selection is also known as "Darwinian" or "directional" selection. See Hurst, "Genetics and the Understanding of Selection," p. 85.

23. T. Bersaglieri et al., "Genetic Signatures of Strong Recent Positive Selection at the Lactase Gene," *American Journal of Human Genetics* 74, no. 6 (2004): 111–20.

24. While farmers were entering central Europe from southeastern Europe 7,500 years ago, milk cows were introduced to the region a bit later as part of what prehistorian Andrew Sherratt (1946–2006) called the "Secondary Products Revolution," which introduced plants and animals that not only had to be slaughtered to be eaten, but that could provide food (such as milk) while being kept alive. The Secondary Products Revolution also introduced alcoholic beverages and the cultural behaviors associated with them. See Sherratt's characteristically enjoyable essay "Cups That Cheered: The Intro-

duction of Alcohol into Prehistoric Europe," in A. Sherratt, *Economy and Society in Prehistoric Europe: Changing Perspectives* (Princeton, NJ: Princeton University Press, 1997), pp. 376–402.

25. S. A. Tishkoff et al., "Convergent Adaptation of Human Lactase Persistence in Africa and Europe," *Nature Genetics* 39, no. 1 (2007): 31–40.

26. Africans were domesticating and herding cattle, including milk cows, as early as five thousand years ago. See M. A. McDonald, "Early African Pastoralism: View from Dakhleh Oasis (South Central Egypt)," *Journal of Anthropological Archaeology* 17, no. 2 (1998): 124–42.

27. Very confusingly, balancing selection is also known as "centrifugal" or "disruptive" selection. See Hurst, "Genetics and the Understanding of Selection," p. 85.

28. Vieillot was also first to scientifically describe the wild turkey, pintail, cedar waxwing, and many other North American birds among dozens of others worldwide, based on his extensive travels. Like many other naturalists of the time, despite his tremendous contribution to Western knowledge, he died largely forgotten and in poverty. See P. H. Oehser, "Louis Jean Pierre Vieillot (1748–1831)," *Auk* 65 (1948): 568–76.

29. T. B. Smith, "Disruptive Selection and the Genetic Basis of Bill Size Polymorphism in the African Finch *Pyrenestes*," *Nature* 363 (1993): 618–20. Research continues in this species because it's one of the few well-documented cases of balancing selection: UCLA's Institute of Tropical Research website states: "We are currently using molecular genetic approaches to examine the population structure and phylogenetic histories of bill [beak] morphs [shapes and sizes] in order to understand how disruptive [balancing] selection may lead to speciation and we are exploring which genes are involved in generating bill [beak] variation." See http://www.ioe.ucla.edu/ctr/research/polymorphisms.html (accessed June 10, 2010).

30. M. R. Gross, "Disruptive Selection for Alternative Life Histories in Salmon," *Nature* 313 (1985): 47–48. A more recent study demonstrated that body size is strongly genetically determined in the Coho; see J. T. Silverstein and W. K. Hershberger, "Genetic Parameters of Size Pre- and Post-

Smoltification in Coho Salmon (*Oncorhynchus kisutch*)," *Aquaculture* 128, no. 1 (1994): 67–77, http://www.sciencedirect.com/science/journal/00448486 (accessed March 10, 2011).

31. A recent review of balancing (or *disruptive*) selection concludes that it has "regained a prominent role in evolutionary thinking, especially in speciation research." See p. 243 in C. Reuffler et al., "Disruptive Selection and Then What?" *Trends in Ecology and Evolution* 21, no. 5 (2006): 238–45. For more on the role of this kind of selection, see chapter 5 of this book.

32. For an excellent review of sexual selection leading to female competition, see T. Clutton-Brock, "Sexual Selection in Females," *Animal Behavior* 77, no. 1 (2009): 3–11.

33. S. J. Hodge et al., "Determinants of Reproductive Success in Dominant Female Meerkats," *Journal of Animal Ecology* 77 (2008): 92–102.

34. Clutton-Brock defines sexual selection, technically, as "a process operating through intrasexual competition for reproductive opportunities, providing a conceptual framework that is capable of incorporating the processes leading to the evolution of secondary sexual characters in both sexes." See p. 1885 in T. Clutton-Brock, "Sexual Selection in Males and Females," *Science* 318 (2007): 1882–85.

35. The quotation is from pp. 114–15 of W. Beebe, "Notes on the Hercules Beetle *Dynastes hercules* (Linn.), at Rancho Grande, Venezuela, with Special Reference to Combat Behavior," *Zoologica* 32 (1947): 109–16.

36. My list is adapted from of E. O. Wilson, *Sociobiology: The New Synthesis* (Cambridge, MA: Belknap/Harvard University Press, 1975), p. 322, table 15-1.

37. Canadian psychologist Merlin Donald has characterized the small-space, short-term "bubble of perception" of nonhuman minds as "episodic," that is, concerned with one episode after another, again and again through the course of life, with (compared to humans) little conception of distant places (small-space) or distant futures or pasts (short-term). See M. Donald, *Origins of the Modern Mind: Three Stages in the Evolution of Culture and Cognition* (Cambridge, MA: Harvard University Press, 1991).

38. Endler, *Natural Selection in the Wild*, p. 23.

39. For an investigation that shows that we might need to reassess our conception of "cloning" in nature—essentially arguing that cloning might be the exception rather than the rule—see J. M. Smith et al., "How Clonal Are Bacteria?" *Proceedings of the National Academy of Sciences of the United States of America* 90 (1993): 4384–88.

40. R. Dawkins, *The Selfish Gene* (Oxford: Oxford University Press, 1976).

41. G. C. Williams, *Natural Selection: Domains, Levels, and Challenges* (New York: Oxford University Press, 1992), pp. 23–37.

42. Late philosopher of evolution David Hull (1935–2010) introduced the concept of the replicators, interactors, and evolving lineages in a fascinating 1980 paper titled "Individuality and Selection," *Annual Review of Ecology and Systematics* 11 (1980): 311–32.

43. F. J. Ayala, *Population and Evolutionary Genetics: A Primer* (Menlo Park, CA: Benjamin Cummings, 1982), pp. 30–31.

44. The literature on group selection is vast and deeply divided. A recent review titled "Group Selection Is Dead! Long Live Group Selection!" points out that one "definition of group selection . . . 'the evolution of traits based on the differential survival and reproduction of groups' . . . does not differ empirically from the similarly broad definition of kin selection: 'selection affected by relatedness among individuals.'" See p. 574 in A. Shavit and R. L. Millstein, "Group Selection Is Dead! Long Live Group Selection!" *BioScience* 58, no. 7 (2008): 574–75.

45. Species selection is as hotly debated as group selection. A not-so-recent review indicates that the jury is still out: see T. A. Grantham, "Hierarchical Approaches to Macroevolution: Recent Work on Species Selection and the 'Effect Hypothesis,'" *Annual Review of Ecology and Systematics* 26 (1995): 301–21. A 2002 review suggests that while it is plausible, species selection has not yet been conclusively demonstrated to everyone's satisfaction; see B. J. Crespi, "Species Selection," in *Encyclopedia of the Life Sciences*, vol. 17 (London: Nature Publishing Group, 2002), pp. 458–561.

46. E. Vrba, "Mammals as a Key to Evolutionary Theory," *Journal of Mammalogy* 73, no. 1 (1992): 1–28, E. S. Vrba, "On the Connections between Palaeoclimate and Evolution," in *Palaeoclimate and Evolution with Emphasis on Human Origins*, ed. E. S. Vrba et al. (New Haven, CT: Yale University Press, 1995), pp. 24–45, and, more recently, E. Vrba, "Mass Turnover and Heterochrony Events in Response to Physical Change," *Paleobiology* 31 (2005): 157–74. For a counterview that reanalyzes some of Vrba's earlier case studies, see R. A. Kerr, "Evolution: New Mammal Data Challenge Evolutionary Pulse Theory," *Science* 273, no. 5274 (1996): 431–32.

47. Ayala, *Population and Evolutionary Genetics*, p. 30.

48. This figure is adapted from figure 2.1 in ibid., p. 31. The population numbers in this example are similar to those reported from a recent study of sand lizards in southern Sweden, where local populations average about two hundred individuals, although there is some significant variation from this figure. See T. Madsen et al., "Population and Genetic Diversity in Sand Lizards (*Lacerta agilis*) and Adders (*Vipera berus*)," *Biological Conservation*, no. 94 (2000): 257–62.

49. Not only are there different levels of biological organization—and evolution "works" on them in different ways—but how evolution works on them might well have changed over time. In *Darwinian Dynamics: Evolutionary Transitions in Fitness and Individuality* (Princeton, NJ: Princeton University Press, 1999), p. 7, Richard E. Michod argues that evolution worked somewhat differently when multicellular organisms arose and when those organisms began to exist in communities rather than in essential isolation. Note—once again—that Michod isn't arguing that basic Darwinian evolution by replication, variation, and selection doesn't happen; he is saying that this evolution may have worked in different ways at different times.

50. Gould is quoted in R. Dawkins, *The Extended Phenotype: The Gene as the Unit of Selection* (San Francisco: W. H. Freeman, 1992), p. 116.

51. Endler, *Natural Selection in the Wild*, p. 163.

52. My table is based on a consideration of many sources; a clear discussion is available in J. M. Savage, *Evolution*, 3rd ed. (New York: Holt, Rinehart

and Winston, 1977), and though it is more than thirty years old, it remains useful.

53. See G. J. Vermeij, *Biogeography and Adaptation* (Cambridge, MA, Harvard University Press, 1978).

CHAPTER 5: THE FACT OF SPECIATION

1. *Lucretius: On the Nature of the Universe*, trans. and ed. R. E. Latham (London: Penguin Books), 1982.

2. See "species," in J. A. Sampson and E. S. C. Weiner, eds., *Oxford English Dictionary*, vol. 16 (Oxford: Clarendon Press, 1989), p. 155.

3. *The Historie of Serpents* was one of several bestiaries written by English cleric Edward Topsell (1572–1625). You can read some of Topsell's works online at the wonderful *Early English Books Online* site, which contains "digital facsimile page images of virtually every work printed in England, Ireland, Scotland, Wales and British North America and works in English printed elsewhere from 1473–1700," at http://eebo.chadwyck.com/home. Go to the site, search for "Topsell," and you're on your way. The University of Houston's *History of Four-Footed Beasts and Serpents* website maintains a gallery of images from Topsell's books at http://info.lib.uh.edu/sca/digital/beast/index.html (accessed March 10, 2011).

4. My quotations from the King James Bible (first printed in 1611) come from the University of Michigan's online edition at http://quod.lib.umich.edu/k/kjv/ (accessed March 10, 2011).

5. Even before Darwin, some naturalists suggested that species might change, countering the mainstream view that they could not change, but these other ideas did not survive after Darwin's theory. A history of the concept of evolution up to 1989 is available in P. J. Bowler, *Evolution: The History of an Idea*, 4th ed. (Berkeley: University of California Press, 1989), but note that there have been big changes since the 1990s; these are introduced in chap. 8 of this book.

6. Richard Dawkins clearly identifies the concept of individuality of life-forms, noting that they (a) cannot be divided without losing their organization and function, (b) have a unique genetic code, and (c) are "walled off" from their peers in one way or another. See R. Dawkins, *The Extended Phenotype: The Gene as the Unit of Selection* (San Francisco: W. H. Freeman, 1982), pp. 250–51.

7. See p. 6606 in K. de Queiroz, "Ernst Mayr and the Modern Concept of Species," *Proceedings of the National Academy of Sciences of the USA* 102, suppl. 1 (2005): 6600–6607.

8. Mayr is quoted on page 6600 of ibid.

9. In 1942, Mayr was refining concepts previously suggested by the biologist Theodosius Dobzhansky (1900–1975), who had written about "reproductive isolation" in his 1937 book *Genetics and the Origin of Species*.

10. Sampson and Weiner, *Oxford English Dictionary*, p. 156.

11. See, for example, various papers in J. Hey, W. M. Fitch, and F. Ayala, eds., *Systematics and the Origin of Species on Ernst Mayr's 100th Anniversary* (Washington, DC: National Academies Press, 2005), the comprehensive treatment of which is in J. A. Coyne and H. A. Orr, *Speciation* (Sunderland, MA: Sinauer, 2004), and the overview of which is in T. E. Wood and L. H. Rieseberg, "Speciation: Introduction," in *Encyclopedia of the Life Sciences*, vol. 17 (London: Nature Publishing Group, 2002), pp. 415–22. A 1995 review that suggests some revision of Mayr's concept—but does not in any way overturn it—is found in J. A. Mallet, "Species Definition for the Modern Synthesis," *Trends in Ecology and Evolution* 10 (1995): 294–99.

12. This isn't just the pronouncement of science: studies with traditional people of New Guinea (among others) have repeatedly found that scientific and folk classifications of life-forms (plants, birds, mammals, and so on) are largely the same. See Coyne and Orr, *Speciation*, pp. 12–13. Even the renegade microbiologist Lynn Margulis—whom we'll meet in chapter 8 of this book and who has little use for many modern concepts in biology—agrees that Mayr's species concept is useful for most plants and animals. See L. Margulis and D. Sagan, *Acquiring Genomes: A Theory of the Origins of Species* (New York: Basic Books, 2002), p. 58.

13. K. de Queiroz, "Ernst Mayr and the Modern Concept of Species," in *Systematics and the Origin of Species: On Ernst Mayr's 100th Anniversary*, ed. J. Hey, W. M. Fitch, and F. Ayala (Washington, DC: National Academies Press, 2005), pp. 243–66, table 13.1.

14. See p. 276 in E. Mayr, "What Is a Species and What Is Not? *Philosophy of Science* 63 (1996): 262–77.

15. See R. M. May, "How Many Species?" *Philosophical Transactions of the Royal Society of London* B, 330 (1990): 293–304. In 2005, biologist E. O. Wilson estimated that there are probably about fifteen to twenty million species of living things; see p. 3 in E. O. Wilson, "Introductory Essay: Systematics and the Future of Biology," in *Systematics and the Origin of Species: On Ernst Mayr's 100th Anniversary*, ed. J. Hey, W. M. Fitch, and F. Ayala, pp. 1–8.

16. It has been calculated that humans drive extinction at a rate unseen for more than two hundred million years; see the detailed discussion of the "sixth extinction" in N. Eldredge, "Cretaceous Meteor Showers, the Human Ecological "Niche," and the Sixth Extinction," in *Extinctions in Near Time: Causes, Contexts, and Consequences*, ed. R. D. E. MacPhee (New York: Kluwer Academic/Plenum, 1999), pp. 1–14, and the more detailed examination in R. MacPhee and C. Flemming, "Requiem Aeternum: The Last Five Hundred Years of Mammalian Species Extinctions," in *Extinctions in Near Time*, ed. R. MacPhee, pp. 333–72. For a more general account, see R. Leakey and R. Lewin, *The Sixth Extinction: Biodiversity and Its Survival* (London: Weidenfeld & Nicolson, 1996).

17. Note that the word *stage* itself is a little problematic, connoting that the process is going somewhere intended, but only with human oversight is the history of a species classified into "stages" (for example, early, middle, and late). For the quotation, see S. J. Gould and N. Eldredge, eds., *Genetics and the Origin of Species by Theodosius Dobzhansky*, Columbia Classics in Evolution Series (New York: Columbia University Press, 1982), p. 312. Regarding the complexities of classifying living things in general, Ayala and Coluzzi write that "Dobzhansky was well aware that complex concepts such as species cannot satisfactorily be defined by a particular set of words encompassed in a single sen-

tence." See p. 48 in F. Ayala and M. Coluzzi, "Chromosome Speciation: Humans, *Drosophila,* and Mosquitos," in *Systematics and the Origin of Species: On Ernst Mayr's 100th Anniversary*, ed. J. Hey, W. M. Fitch, and F. Ayala, pp. 46–68.

18. Margulis and Sagan, *Acquiring Genomes*, pp. 15–16.

19. The misconception of Nature (with a capital N) as a smoothly functioning system with jobs to be filled by certain species is dispelled in C. M. Smith and C. Sullivan, *The Top Ten Myths about Evolution*, chap. 7 (Amherst, NY: Prometheus Books, 2006).

20. This definition is from p. 415 of T. E. Wood and L. H. Rieseberg, *Speciation: Introduction, Encyclopedia of the Life Sciences*, vol. 17 (London: Nature Publishing Group, 2002), pp. 415–22.

21. J. W. Boughman, "Divergent Sexual Selection Enhances Reproductive Isolation in Sticklebacks," *Nature* 411 (2001): 944–48. We'll meet the sticklebacks again later in this book.

22. O. Seehausen, J. J. M. van Alphen, and F. White, "Cichlid Fish Diversity Threatened by Eutrophication That Curbs Sexual Selection," *Science* 277 (1997): 1811.

23. E. C. Minkoff, *Evolutionary Biology* (Reading, MA: Addison-Wesley, 1983).

24. T. Sota and K. Kubota, "Genital Lock-and-Key as a Selective Agent against Hybridization," *Evolution* 52 (1998): 1507–13. A more recent paper using molecular data confirms this case and extends it to the case of other beetle species mechanically isolated by genital shape: see N. Nagata et al., "Historical Divergence of Mechanical Isolation Agents in the Ground Beetle *Carabus arrowianus* as Revealed by Phylogeographical Analyses," *Molecular Ecology* 18, no. 7 (2009): 1408–21.

25. For the coral species examples, see N. Knowlton et al., "Direct Evidence for Reproductive Isolation among the Three Species of the *Montastraea annularis* Complex in Central America (Panamá and Honduras)," *Marine Biology* 127, no. 4 (1997): 705–11. Note that temporal isolation, though not as widely studied as some other modes of reproductive isolation, remains theo-

retically likely to have had a "pivotal" role in many speciations: see Coyne and Orr, *Speciation*, pp. 204–10.

26. Wood and Rieseberg, "Speciation: Introduction," in *Encyclopedia of the Life Sciences*, pp. 415–22, and E. Bakker, *An Island Called California* (Berkeley: University of California Press, 1984).

27. N. Aspinwall, "Genetic Analysis of North American Populations of the Pink Salmon, *Oncorhyncus gorbuscha*, Possible Evidence for the Neutral Mutation-Random Drift Hypothesis," *Evolution* 28 (1974): 295–305.

28. For more on these other forms of postmating isolation, see Coyne and Orr, *Speciation*, pp. 232–46.

29. M. Schilthuizen, *Frogs, Flies, and Dandelions—Speciation, The Origin of New Species* (Oxford: Oxford University Press, 2001), pp. 177–78.

30. See p. 255 in D. J. Futuyama and G. C. Mayer, "Non-Allopatric Speciation in Animals," *Systematic Zoology* 29, no. 3 (1980): 254–71.

31. The "genetic clock" is based on the fact that genomes (the DNA of a species) accumulate mutations at largely known and largely stable rates. Comparing the genomes of two hypothetically related species (such as humans and chimpanzees, for example) shows the number of differences between those genomes, giving an estimate of how much time has elapsed since they were, in effect, the same species. The method has its flaws, but these are well-known, and techniques are constantly refined. A technical review can be found in T. G. Barraclough and S. Nee, "Phylogenetics and Speciation," *TRENDS in Ecology and Evolution* 16, no. 7 (2001): 391–99 (especially box 1 on p. 393); a more general overview is presented in L. Bromham and D. Penny, "The Modern Molecular Clock," *Nature Reviews Genetics* 4, no. 3 (2003): 216–24.

32. N. Knowlton et al., "Divergence in Proteins, Mitochondrial DNA, and Reproductive Compatibility across the Isthmus of Panama," *Science* 260 (1993): 1629–32. A recent review of many other species on either side of the isthmus of Panama backs up—and provides details regarding—this study; see H. A. Lessios, "Great American Schism: Divergence of Marine Organisms after the Rise of the Central American Isthmus," *Annual Review of Ecology, Evolution, and Systematics* 39 (2008): 63–91.

33. This is one of many cases: see H. L. Carson and A. R. Templeton, "Genetic Revolutions in Relation to Speciation Phenomena: The Founding of New Populations," *Annual Review of Ecology and Systematics* 15 (1984): 97–131.

34. A. Sato et al., "On the Origin of Darwin's Finches," *Molecular Biology and Evolution* 18, no. 3 (2000): 299–311.

35. You can read the remarkable story of the Grants' long-term studies in J. Weiner, *The Beak of the Finch: A Story of Evolution in Our Time* (New York: Vintage/Random House, 1995). A scientific report is available in P. R. Grant, B. R. Grant, and K. Petren, "The Allopatric Phase of Speciation: The Sharp-beaked Ground Finch (*Geospiza difficilis*) on the Galapagos Islands," *Biological Journal of the Linnean Society* 69 (2000): 287–317.

36. Speciation arising from larger genetically isolated populations is often referred to as *vicariant* speciation, and from smaller populations, *founder effect speciation*, referring to the founder effect of very small initial populations we saw in chapter 3 (it is also known as *peripatric speciation*; *peri* refers to the outer edge, or *peri*phery). For our purposes, these distinctions aren't necessary. A technical review can be found in Coyne and Orr, *Speciation*, chap. 3, and a more general review in Schilthuizen, *Frogs, Flies, and Dandelions*.

37. For original reports, see D. E. Irwin, "Speciation by Distance in a Ring Species," *Science* 307 (2005): 414–16, and D. E. Irwin, "Song Variation in an Avian Ring Species," *Evolution* 54, no. 3 (2000): 998–1010. While this is an interesting case, in a review it has been suggested that more data are needed to cement the case: see Coyne and Orr, *Speciation*, pp. 102–3.

38. Gregory Bateson (1904–1980) is quoted on p. 1 of M. D. White, *Modes of Speciation* (San Francisco: W. H. Freeman, 1978). Bateson's son, Gregory, later became a titan of biology, as revealed in L. E. Bruni, "Gregory Bateson's Relevance to Current Molecular Biology," in *A Legacy for Living Systems: Gregory Bateson as Precursor to Biosemiotics*, vol. 2, ed. J. P. Hoffmeyer (New York: Springer, 2008), pp. 93–119.

39. See p. 295 in T. D. Kocher, "Adaptive Evolution and Explosive Speciation: The Cichlid Fish Model," *Nature Reviews Genetics* 5 (2004): 288–98.

40. When antievolutionists indicate that field observations of speciation are rare, evolutionists counter that we can observe speciation in the laboratory; the antievolutionist's counter to *that* is that the lab is an unnatural setting. But many other biological processes that nobody questions, such as cell division, are also more conveniently observed in a laboratory setting.

41. *Lucretius: On the Nature of the Universe*, p. 107.

42. See Gould's 1982 essay "Evolution as Fact and Theory," online at http://www.harvardsquarelibrary.org/speakout/gould.html (accessed March 10, 2011).

43. M. J. Niemiller, B. J. Fitzpatrick, and B. T. Miller, "Recent Divergence with Gene Flow in Tennessee Cave Salamanders (Plethodontidae: *Gyrinophilus*) Inferred from Gene Genealogies," *Molecular Ecology* 17 (2008): 2258–75.

44. K. Byrne and R. A. Nichols, "*Culex pipiens* in London Underground Tunnels: Differentiation between Surface and Subterranean Populations," *Heredity* 82 (1999): 7–15.

45. J. R. Weinberg, V. R. Starczak, and D. Jörg, "Evidence for Rapid Speciation Following a Founder Effect in the Laboratory," *Evolution* 46, no. 4 (1992): 1214–20.

46. G. Kilias, S. N. Alahiotis, and M. Pelecanos, "A Multifactorial Genetic Investigation of Speciation Theory Using *Drosophila melanogaster*," *Evolution* 34, no. 4 (1980): 730–37.

47. There is an extensive literature on stickleback evolution, which is now as intensively studied as that of the fruit fly, the Galapagos finch, and a number of flower species. One estimate is that about one hundred labs worldwide are focusing on stickleback evolution. A good review with excellent images is available in E. Pennisi, "Changing a Fish's Bony Armor in the Wink of a Gene," *Science* 304 (2004): 1736–39. A more technical paper directly addresses *Pitx1*: see M. D. Shapiro et al., "Genetic and Developmental Basis of Evolutionary Pelvic Reduction in Threespine Sticklebacks," *Nature* 428 (2004): 717–23. More recent papers continue to sharpen focus on *Pitx1* and stickleback evolution. Good overviews of stickleback evolution can be found in H. D. Rundle et

al., "Natural Selection and Parallel Sympatric Speciation in Sticklebacks," *Science* 287 (2000): 306–7, and the well-illustrated report in J. S. McKinnon and H. D. Rundle, "Speciation in Nature: The Threespine Stickleback Model System," *TRENDS in Ecology & Evolution* 17, no. 110 (2002): 480–88.

48. R. B. Langerhans, M. E. Gifford, and E. O. Joseph, "Ecological Speciation in Gambusia Fishes," *Evolution* 61, no. 9 (2007): 2056–74.

49. Earlier reports that Lake Victoria was completely dry around fifteen thousand years ago and was subsequently recolonized when it refilled have been challenged, but DNA studies confirm that whether the lake dried entirely or only partially, Lake Victoria cichlid diversity is a result of adaptive radiation and speciation; see K. R. Elmer et al., "Pleistocene Desiccation in East Africa Bottlenecked but Did Not Extirpate the Adaptive Radiation of Lake Victoria Haplochromine Cichlid Fishes," *Proceedings of the National Academy of Sciences* (USA) 106, no. 32 (2009): 13404–9.

50. R. C. Albertson, J. T. Streelman, and T. D. Kocher, "Genetic Basis of Adaptive Shape Differences in the Cichlid Head," *Journal of Heredity* 94, no. 4 (2003): 291–301. A superb review of cichlid fish evolution is available in T. D. Kocher, "Adaptive Evolution and Explosive Speciation: The Cichlid Fish Model," *Nature Reviews Genetics* 5 (2004): 288–98.

51. M. Barluenga et al., "Sympatric Speciations in Nicaraguan Crater Lake Cichlid Fish," *Nature* 439 (2006): 719–23.

52. J. Ramsey, H. D. Bradshaw Jr., and D. W. Schemske, "Components of Reproductive Isolation between the Monkeyflowers *Mimulus lewisii* and *M. cardinali* (*Phrymaceae*)," *Evolution* 57 (2003): 1520–34, and D. W. Schemske and H. D. Bradshaw Jr., "Pollinator Preference and the Evolution of Floral Traits in Monkeyflowers (*Mimulus*)," *Proceedings of the National Academy of Sciences* (USA) 96 (2003): 11910–15.

53. The Smithsonian Institution's paleobiology webpage on dating methods is a good review: http://paleobiology.si.edu/geotime/main/foundation_dating1.html (accessed March 10, 2011).

54. See U. Sorhannus et al., "Iterative Evolution in the Diatom Genus *Rhizosolenia* (Ehrenberg)," *Lethaia* 24, no. 1 (2001): 39–44.

55. The debate over whether life-forms change gradually (gradual evolution, also known as *gradualism*) or in quick jumps (punctuated evolution, otherwise known as *punctuated equilibrium*) is long; for our purposes it is enough to say that there is evidence that both punctuated and gradual evolution are known in the fossil record and among living species and (most important) that neither model suggests that "evolution is wrong," only that there is debate about just how evolution happens. A good entrance to the debate is found in the 2002 edition of the *Encyclopedia of the Life Sciences* (London: Nature Publishing Group).

56. The fossil record of the genus *Homo* and its closest relatives is covered in many books and articles, and they are hard to keep updated, as new material is discovered continually. A good review, from 2004, is found in R. Lewin and R. Foley, *Principles of Human Evolution* (London: Blackwell Science, 2004). A nice overview of the Laetoli footprints, with a video clip, is available at the Public Broadcasting Service's evolution site at http://www.pbs.org/wgbh/evolution/library/07/1/l_071_03.html (accessed March 10, 2011).

CHAPTER 6: THE FACT OF EVOLUTION

1. *Lucretius: On the Nature of the Universe*, trans. and ed. R. E. Latham (London: Penguin Books, 1982).

CHAPTER 7: EVOLUTION IN ACTION

1. Herodotus, *The Histories*, Book II, English translation by A. D. Godley (Cambridge, MA: Harvard University Press, 1920), p. 68, available online at the Perseus Project website: http://www.perseus.tufts.edu/hopper/ (accessed March 10, 2011).

2. S. P. Doolan and D. W. McDonald, "Breeding and Juvenile Survival among Slender-Tailed Meerkats (*Suricata suricatta*) in the South-Western Kalahari: Ecological and Social Influences," *Journal of Zoology* 242 (1997): 309–27.

3. See p. 877 in R. Gal and F. Libersat, "A Parasitoid Wasp Manipulates the Drive for Walking of Its Cockroach Prey," *Current Biology* 18 (2008): 877–82.

4. T. Spight, "Availability and Use of Shells by Intertidal Hermit Crabs," *Biology Bulletin* 152 (1977): 120–33.

5. W. H. Baltosser, "Nectar Availability and Habitat Selection by Hummingbirds in Guadalupe Canyon," *Wilson Bulletin* 101, no. 4 (1989): 559–78.

6. D. A. Reznick, H. Bryga, and J. A. Endler, "Experimentally Induced Life-History Evolution in a Natural Population," *Nature* 346 (1990): 357–59.

7. Ibid.

8. Ibid.

9. E. Melendez-Ackerman, D. R. Campbell, and N. M. Waser, "Hummingbird Behavior and Mechanisms of Selection on Flower Color in Ipomopsis," *Ecology* 78, no. 8 (1997): 2532–41.

10. L. M. DeVantier, "Rafting of Tropical Marine Organisms on Buoyant Coralla," *Marine Ecology Progress*, Series 86 (1992): 301–2.

11. B. Helmuth, R. R. Veit, and R. Holberton, "Long-Distance Dispersal of a Subantarctic Brooding Bivalve (*Gaimardia trapesina*) by Kelp-Rafting," *Marine Biology* 120 (1994): 421–26.

12. P. Joikel and F. J. Martinelli, "The Vortex Model of Coral Reef Biogeography," *Journal of Biogeography* 19 (1992): 449–58.

13. Based on a recent examination of about a hundred entire genomes, the average number of base pairs in a prokaryote (for example, microscopic life-forms, like bacteria) gene is about 924 base pairs, while the average gene in eukaryotes (for example, multicellular animal life-forms) is about 1,324 base pairs long. See L. Xu et al., "Average Gene Length Is Highly Conserved in Prokaryotes and Eukaryotes and Diverges Only between the Two Kingdoms," *Molecular Biology and Evolution* 23, no. 6 (2006): 1107–8.

14. C. A. Suttle, "Viruses in the Sea," *Nature* 437 (2005): 356–61.

15. G. Jones and E. C. Teeling, "The Evolution of Echolocation in Bats," *Trends in Ecology and Evolution* 21, no. 3 (2006): 149–56.

16. P. T. Madsen et al., "Bisonar Performance of Foraging Beaked Whales

(*Mesoplodon densirostris*)," *Journal of Experimental Biology* 208 (2005): 181–94.

17. Y. Li et al., "The Hearing Gene Prestin Unites Echolocating Bats and Whales," *Current Biology* 20, no. 2 (2008): R55–R56.

18. J. Zheng et al., "Prestin Is the Motor Protein of Cochlear Outer Hair Cells," *Nature* 405 (2000): 149–55, and E. C. Teeling, "Hear, Hear; The Convergent Evolution of Echolocation in Bats," *Trends in Ecology and Evolution* 24, no. 7 (2009): 351–54.

19. G. L. Thomas and R. E. Thorne, "Night-Time Predation by Steller Sea Lions," *Nature* 411 (2001): 1013.

20. E. H. Kovacs and I. Sas, "Cannibalistic Behaviour of *Epidalea* (Bufo) *viridis* Tadpoles in an Urban Breeding Habitat," *North-Western Journal of Zoology* 5, no. 1 (2009): 206–8.

21. S. J. Simpson et al., "Gregarious Behavior in Desert Locusts Is Evoked by Touching Their Back Legs," *Proceedings of the National Academy of Sciences* (USA) 98, no. 7 (2001): 3895–97.

22. M. Hori, "Frequency-Dependent Natural Selection in the Handedness of Scale-Eating Cichlid Fish," *Science* 260 (1993): 216–19.

23. See p. 53 in S. A. West et al., "The Social Lives of Microbes," *Annual Review of Ecology, Evolution, and Systematics* 38 (2007): 53–77. Another interesting review is found in B. J. Crespi, "The Evolution of Social Behavior in Microorganisms," *Trends in Ecology and Evolution* 16, no. 4 (2001): 178–83.

24. One reason, of many, to study such biofilms is that they compose the persistent film that coats the lung tissues of patients of cystic fibrosis. Why study the world of living things, even topics as esoteric as microbial social behavior? First, it is necessary to understand the world we live in. Second, it might well be useful in understanding and solving some of our problems. See J. W. Costerton, "Cystic Fibrosis Pathogenesis and the Role of Biofilms in Persistent Infection," *Trends in Microbiology* 9, no. 2 (2001): 50–52.

25. R. M. Harshey, "Bacterial Motility on a Surface: Many Ways to a Common Goal," *Annual Review of Microbiology* 57 (2003): 249–73.

26. M. L. Porter and K. A. Crandall, "Lost Along the Way: The Signifi-

cance of Evolution in Reverse," *TRENDS in Ecology and Evolution* 18, no. 10 (2003): 541–47.

27. J. B. W. Wolf et al., "Tracing Early Stages of Species Differentiation: Ecological, Morphological, and Genetic Divergence of Galapagos Sea Lion Populations," *BMC Evolutionary Biology* 8 (2008): 150, doi:10.1186/1471-2148-8-150 (accessed March 10, 2011).

28. Darwin, writing in 1859, is quoted on p. 808 of M. Pagel, "Natural Selection 150 Years On," *Nature* 457 (2009): 808–11.

29. The age of the earth was intensely debated throughout the eighteenth and nineteenth centuries. Many calculations used the oldest-known history at the time, the Christian Bible, which suggested that God created earth on the order of six thousand years ago. Roman academic Dionysius Exiguus introduced the split between BC (before Christ) and AD (Anno Domini, or years after Christ). As explorers discovered people and continents (such as Indians living in the Americas) that were not accounted for by the Bible, faith in its use as a perfect history eroded, and as early as 1692 it was argued in Thomas Burnett's "Archaeologicae Philosophiae" that the narrative of Eden and the date of the Creation were fables, not intended to be taken literally. By the 1700s, even though geology was still young, the biblical chronology was severely challenged. By the 1850s, geologists had conclusive evidence that the world was millions—perhaps even billions—of years old. See D. R. Dean, "The Age of the Earth Controversy," *Annals of Science* 38 (1981): 435–56. Recent calculations put the age of the earth at about 4.5 billion (4,500 million) years; see G. B. Dalrymple, *The Age of the Earth* (Stanford: Stanford University Press, 1991).

30. More on Paley's argument from design—which has been dismantled many times—is available in C. M. Smith and C. Sullivan, *The Top Ten Myths about Evolution* (Amherst, NY: Prometheus Books, 2006), chap. 9.

31. See p. 59 (lines 865–70) of *Lucretius: On the Nature of the Universe*, trans. and ed. R. E. Latham (London: Penguin Books, 1982).

32. Although in the opening of his text "On the Nature of the Universe" (sometimes known as "On the Nature of Things") Lucretius apparently

thanked the gods for his own powers of perception, it is clear that this was done for convention; later in the work he clearly states that "nature is free and uncontrolled by proud masters and runs the universe by herself without the aid of gods." See p. 64 (lines 1092–1093) of *Lucretius: On the Nature of the Universe*. Lucretius's understanding of even the foundations of evolution—replication, variation, and selection are all described in detail in his text—came just shy of building them into the principles of Darwinian evolution. This was recognized as early as 1874, when physicist and naturalist John Tyndall (1820–1893) showed that Paley's argument had been shown to be flawed by Lucretius's rejection of "intelligent design"; see J. Tyndall, *Address Delivered Before the British Association Assembled at Belfast, with Additions* (London: Longmans and Green, 1874), p. 8.

33. Outside of science, vitalism is still influential, and in fact most people worldwide believe that supernatural forces are involved in life processes. This consensus is based not on scientific evidence but on cultural tradition, including myth and religion, both examined in chapter 9. The wide consensus of vitalism is not evidence of its being true: before microscopy discovered that germs spread disease, everyone on Earth was wrong about how disease spreads. Science is not immune to being wrong by consensus, but it has the self-correcting mechanisms of demand for evidence and experimentation that are absent from—indeed actively suppressed by—religious systems of knowledge based on authority of ancient texts considered to be inerrant.

34. E. Sober, *Evidence and Evolution* (Cambridge: Cambridge University Press, 2008), pp. 161–62.

35. For a review of how bacteria assemble flagella by natural rather than by supernatural means, see R. Macnab, "How Bacteria Assemble Flagella," *Annual Review of Microbiology* 57 (2003): 77–100.

36. Aristotle, "Parts of Animals," online at http://classics.mit.edu/Aristotle/parts_animals.html (accessed March 10, 2011).

CHAPTER 8: THE MIRROR-HOUSE OF EVOLUTION

1. See p. 173 in C. Woese, "A New Biology for a New Century," *Microbiology and Molecular Biology Reviews* (June 2004): 173–86.

2. For a flavor of the times, see J. Huxley, *Evolution: The Modern Synthesis* (London: Allen & Unwin, 1942). For a more recent history of evolutionary thought, see C. Zimmer, *Evolution: The Triumph of an Idea* (New York: HarperCollins, 2001).

3. See p. 474 in E. V. Koonin, "*The Origin* at 150: Is a New Evolutionary Synthesis in Sight?" *Trends in Genetics* 25, no. 11 (2009): 473–75.

4. For a historical review of the "modern" synthesis, see E. Mayr and W. B. Provine, eds., *The Evolutionary Synthesis: Perspectives on the Unification of Biology,* with a new preface by Ernst Mayr (Cambridge, MA: Harvard University Press, 1998). For "coverage" of the current "revolution," see N. Goldenfeld and C. Woese, "Biology's Next Revolution," *Nature* 445 (2007): 369, M. R. Rose and T. H. Oakley, "The New Biology: Beyond the Modern Synthesis," *Biology Direct* 2 (2007), doi:10.1186/1745-6150-2-30 (accessed March 10, 2011), Koonin, "*The Origin* at 150," pp. 473–75, and E. V. Koonin, "Darwinian Evolution in the Light of Genomics," *Nucleic Acids Research* 37, no. 4 (2009): 1011–34.

5. M. R. Rose and T. H. Oakley, "The New Biology: Beyond the Modern Synthesis," *Biology Direct* 2 (2007): p. 1, doi:10.1186/1745-6150-2-30 (accessed March 10, 2011).

6. See p. 183 in Woese, "A New Biology for a New Century."

7. A good review of the findings of the DNA clock can be found in A. R. Templeton, "Out of Africa Again and Again," *Nature* 416 (2002): 45–51.

8. See pp. 97–99 in B. Sykes, "Using Genes to Map Population Structure and Origins," in *The Human Inheritance*, ed. B. Sykes (Oxford: Oxford University Press, 1999), pp. 93–117.

9. See p. 48 in J. H. Paul, "Microbial Gene Transfer: An Ecological Perspective," *Journal of Molecular Biotechnology* 1, no. 1 (1999): 45–50.

10. See p. 123 in G. Myers, I. Paulsen, and C. Fraser, "The Role of Mobile

DNA in the Evolution of Prokaryotic Genomes," in *The Implicit Genome*, ed. L. H. Caporale (New York: Oxford University Press, 2006), pp. 121–37.

11. Ibid.

12. Ibid.

13. See p. 270 in A. B. Simonson et al., "Decoding the Genomic Tree of Life," in *Systematics and the Origin of Species: On Ernst Mayr's 100th Anniversary*, ed. J. Hey, W. M. Fitch, and F. Ayala (Washington, DC: National Academies Press, 2005), pp. 267–85.

14. See p. 993 in G. J. Olsen and C. Woese, "Archaeal Genomics: An Overview," *Cell* 89 (1997): 991–94.

15. See p. 125 in J. R. Brown, "Ancient Horizontal Gene Transfer," *Nature Reviews Genetics* 4 (2003): 121–32. This paper is an excellent review of the issue.

16. See p. 6 in T. R. E. Southwood, "Interactions of Plants and Animals: Patterns and Processes," *Oikos* 44 (1985): 5–11.

17. See p. 611 in P. J. Keeling and J. D. Palmer, "Horizontal Gene Transfer in Eukaryotic Evolution," *Nature Reviews Genetics* 9 (2008): 605–18, in which the authors also note (p. 605) that "the number of well-supported cases of transfer from both prokaryotes and eukaryotes is now expanding rapidly." Prokaryotes are microscopic life-forms without a wall around the nucleus (which contains the DNA), whereas eukaryotes have a wall around the nucleus; animal life is eukaryotic. Note that there is plenty of debate about the significance (or reality) of some of the terminology here. The great microbiologist Carl Woese makes his feelings known by writing in 2004 that in 1962 "the term (and concept) 'prokaryote' slithered onto the scene." See p. 177 in Woese, "A New Biology for a New Century."

18. Lamarck's ideas are sometimes oversimplified, but here I think the essence is clear enough. You can read more about Lamarck and how his ideas differed from Darwin's in C. M. Smith and C. Sullivan, *The Top Ten Myths about Evolution* (Amherst, NY: Prometheus Books, 2006), chap. 2.

19. See E. Bapteste and Y. Boucher, "Lateral Gene Transfer Challenges Principles of Microbial Systematics," *Trends in Microbiology* 16, no. 5 (2008):

200–207, fig. 3, and E. Bapteste et al., "Phylogenetic Reconstruction and Lateral Gene Transfer," *Trends in Microbiology* 12, no. 9 (2004): 406–11, fig. 4, for new methods of visualizing the true nature of species and their relationships.

20. E. V. Koonin and Y. Boucher, "Is Evolution Darwinian or/and Lamarckian?" *Biology Direct* 4 (2009), doi:10.1186/1745-6150-4-42 (accessed March 10, 2011).

21. M. B. Gogarten, J. P. Gogarten, and L. Oldenzewski, *Horizontal Gene Transfer: Genomes in Flux* (New York: Humana Press, 2009). Shorter reviews are available in J. P. Gogarten and F. Townsend, "Horizontal Gene Transfer, Genome Innovation and Evolution," *Nature Reviews Microbiology* 3 (2005): 679–87, and J. P. Gogarten, W. F. Doolittle, and J. G. Lawrence, "Prokaryotic Evolution in Light of Gene Transfer," *Molecular Biological Evolution* 19, no. 12 (2002): 2226–38.

22. See p. 122 in G. Myers, I. Paulsen, and C. Fraser, "The Role of Mobile DNA in the Evolution of Prokaryotic Genomes," in *The Implicit Genome*, ed. L. H. Caporale (New York: Oxford University Press, 2006), pp. 121–37.

23. E. V. Koonin, "Darwinian Evolution in the Light of Genomics," *Nucleic Acids Research* 37, no. 4 (2009): 1011–34.

24. For an overview of phenotypic plasticity, see A. A. Agrawal, "Phenotypic Plasticity in the Interactions and Evolution of Species," *Science* 294 (2001): 321–26, or a more technical and thorough treatment in T. J. Dewitt and S. M. Scheiner, *Phenotypic Plasticity: Functional and Conceptual Approaches* (Oxford: Oxford University Press, 2004), in which many examples of phenotypic plasticity are presented, and in which plasticity is referred to as being "activated" by "environmental cues"—general ideas worth considering when we view or imagine the real life of any life-form in its selective environment.

25. R. Levins, *Evolution in Changing Environments* (Princeton, NJ: Princeton University Press, 1968).

26. K. Parjeko, "Embryology of *Chaoborus*-Induced Spines in *Daphnia pulex*," *Hydrobiologia* 231 (1992): 77–84.

27. See p. 73 in M. J. West-Eberhard, "Developmental Plasticity and the

Origin of Species," in *Systematics and the Origin of Species on Ernst Mayr's 100th Anniversary*, ed. J. Hey, W. M. Fitch, and F. Ayala (Washington, DC: National Academies Press, 2005), pp. 69–94.

28. S. C. Stearns, "The Evolutionary Significance of Phenotypic Plasticity," *Bioscience* 37, no. 7 (1989): 436–45.

29. Biologist Miquel Porta, noting that the relationship between genes and what they build is not always as direct as we initially expected, has characterized the genome not as a rigid computer program but more like a "jazz score." See M. Porta, "The Genome Sequence Is a Jazz Score," *International Journal of Epidemiology* 32, no. 1 (2003): 29–31.

30. R. E. Green et al., "Analysis of One Million Base Pairs of Neanderthal DNA," *Nature* 444 (2006): 330–36, R. E. Green et al., "A Draft Sequence of the Neandertal Genome," *Science* 328, no. 5979 (2010): 710–22, and Pete Spotts, "Cavemen among Us: Some Humans Are 4 Percent Neanderthal," *Christian Science Monitor*, May 6, 2010, http://www.csmonitor.com/Science/2010/0506/Cavemen-among-us-Some-humans-are-4-percent-Neanderthal (accessed March 10, 2011).

31. Jennifer Vargas, "Fossilized Eggshells Yield DNA," *Discovery News*, March 9, 2010.

32. G. A. Logan, J. J. Boon, and G. Eglinton, "Structural Biopolymer Preservation in Miocene Leaf Fossils from the Clarkia Site, Northern Idaho," *Proceedings of the National Academy of Sciences USA* 90 (1993): 2246–50.

33. R. DeSalle et al., "DNA Sequences from a Fossil Termite in Oligo-Miocene Amber and Their Phylogenetic Implications," *Science* 257 (1992): 1933–36.

34. John Roach, "Oldest Dinosaur Protein Found—Blood Vessels, More," *National Geographic News*, May 1, 2009, http://news.nationalgeographic.com/news/2009/05/090501-oldest-dinosaur-proteins.html (accessed September 30, 2010).

35. S. Pääbo, "Ancient DNA," in *DNA: Changing Science and Society*, ed. T. Krude (Cambridge: Cambridge University Press, 2004), pp. 68–87.

36. E. S. Witkin, "Ultraviolet Mutagenesis and the SOS Response in

Escherichia coli: A Personal Perspective," *Environmental and Molecular Mutagenesis* 14, suppl. 16 (1989): 30–34.

37. See p. 39 in E. C. Friedberg, "Mutation as a Phenotype," in *The Implicit Genome*, ed. L. H. Caporale, pp. 39–56.

38. Ibid., pp. 39–56, Friedberg, Walker, and Seide, "DNA Repair and Mutagenesis," and several articles in the December 23, 1994, issue of *Science*.

39. Boveri is quoted in F. Theodor Boveri Baltzer, *The Life of a Great Biologist*, translated from the German by Dorothea Rudnick (Berkeley: University of California Press, 1967). The quotation can be found at Sinauer Publishing's Developmental Biology textbook site at http://8e.devbio.com/article.php?ch=7&id=75 (accessed March 10, 2011).

40. S. B. Carroll, *Endless Forms Most Beautiful: The New Science of Evo-Devo and the Making of the Animal Kingdom* (New York: W. W. Norton, 2005), p. 577. A shorter and more technical review is found in S. B. Carroll, "Evo-Devo and an Expanding Evolutionary Synthesis: A Genetic Theory of Morphological Evolution," *Cell* 134 (2005): 25–36.

41. Note that, as usual, things are not simple in the world of living systems; while *pax6* is important to the development of photoreceptors, other genes are also involved in the development of the eight main kinds of animal eyes. See R. D. Fernald, "Casting a Genetic Light on the Evolution of Eyes," *Science* 313 (2006): 1914–18. Genes that control multiple characteristics of the phenotype (the organism built by the genes) are called *pleiotropic* (*pleio* referring to "many," and *tropic* referring to "ways"), and single characteristics—like the color of the skin—that are controlled by multiple genes are called *polygenic*. Imagine the complexity!

42. The first heart structures are known from about five hundred million years ago; see E. N. Olson, "Gene Regulatory Networks in the Evolution and Development of the Heart," *Science* 313 (2006): 1922–27.

43. Y.-Y. Shen et al., "Parallel and Convergent Evolution of the Dim-Light Vision Gene *RH1* in Bats (Order: Chiroptera)," *PLoS Biology* 15, no. 1 (2010), doi:10.1371/journal.pone.0008838 (accessed March 10, 2011).

44. There is a vast literature on developmental evolutionary biology

today, and though I have surveyed some of it, your best guide is found in the beautifully illustrated book by Sean Carroll, *Endless Forms Most Beautiful.*

45. See pp. 366–67 in C. R. Marshall, "Explaining the Cambrian 'Explosion' of Animals," *Annual Review of Earth and Planetary Sciences* 34 (2006): 355–84.

46. J. N. Thompson, "The Evolution of Species Interactions," *Science* 284 (1999): 2116–18, for the quotation. For another review emphasizing that coevolution plays a central role in evolution, see J. N. Thompson, S. L. Nuismer, and R. Gomulkiewicz, "Coevolution and Maladaptation," *Integrative and Comparative Biology* 42 (2002): 381–87. An older but still good review is found in G. J. Vermeij, "The Evolutionary Interaction among Species: Selection, Escalation, and Coevolution," *Annual Review of Ecology and Systematics* 25 (1994): 219–36.

47. L. Margulis and D. Sagan, *Acquiring Genomes: A Theory of the Origins of Species* (New York: Basic Books, 2002), pp. 176–79.

48. See p. 730 in I. Zillber-Rosenberg and E. Rosenberg, "Role of Microorganisms in the Evolution of Animals and Plants: The Hologenome Theory of Evolution," *FEMS Microbiology Review* 32 (2008): 723–35.

49. For a current and thorough overview of symbiosis, see A. E. Douglas, *The Symbiotic Habit* (Princeton, NJ: Princeton University Press, 2010).

50. See p. 723 in Zillber-Rosenberg and Rosenberg, "Role of Microorganisms in the Evolution of Animals and Plants.

51. E. Mayr, foreword to *Acquiring Genomes: A Theory of the Origins of Species,* by L. Margulis and D. Sagan (New York: Basic Books, 2002).

52. Ibid., p. 2.

53. Williamson is quoted in ibid., p. 166.

54. Koonin, "*The Origin* at 150," pp. 473–75, table 1.

55. The full list is available at http://www.fas.harvard.edu/~fqeb/questions/. Plenty of other scientific research efforts are addressing problems just as interesting as these.

56. See p. 179 in Woese, "A New Biology for a New Century." An excellent review of the last decade of genetic and genomic studies summarized by

eight leading researchers in these fields is found in E. Heard et al., "Ten Years of Genetics and Genomics: What Have We Achieved and Where Are We Heading?" *Nature Reviews Genetics* (2010), advance online publication at doi:10.1038/nrg2878 (accessed March 10, 2011).

57. See p. 173 in Woese, "A New Biology for a New Century."

CHAPTER 9: THE GRAND ILLUSION

1. E. S. Donno, ed., *Andrew Marvell: The Complete Poems* (London: Penguin, 1972), p. 101.

2. The earliest symbolic artifacts widely accepted by archaeologists are the geometrically incised stones found at Blombos Cave, South Africa, dated to well over seventy-five thousand years ago; see C. S. Henshilwood et al., "Emergence of Modern Human Behavior: Middle Stone Age Engravings from South Africa," *Science* 295, no. 5558 (2002): 1278–80.

3. See p. 526 in D. Pilbeam's afterword to *Neanderthals and Modern Humans in Western Asia*, ed. T. Akazawa, K. Aoki, and O. Bar-Yosef (New York: Plenum Press, 1998), pp. 523–27.

4. D. L. Cheney and R. M. Seyfarth, "Social and Non-Social Knowledge in Vervet Monkeys," *Philosophical Transactions of the Royal Society B* 308 (1985): 187–201.

5. How the mind (what the physical brain does) actually evolved (changed through time from an ancestral to a modern form) is a fascinating study, but there isn't room for it in this book. The *Stanford Encyclopedia of Philosophy* warns that "[p]erhaps no aspect of mind is more familiar or more puzzling than consciousness and our conscious experience of self and world." E. N. Zalta, ed., *Stanford Encyclopedia of Philosophy*, August 2004, http://plato .stanford.edu/entries/consciousness/. Both Merlin Donald (in his 1993 book *Origins of the Modern Mind: Three Stages in the Evolution of Culture and Cognition*) and Steven Mithen (in his 1999 book *The Prehistory of the Mind*) have made significant progress on the issue. An overview of their approaches and

conclusions is found in C. M. Smith, "Rise of the Modern Mind," *Scientific American MIND*, August 2006, pp. 73–79.

6. The concept that active creation is central to human cognition is not mine; I have pulled it into my conception from L. Gabora, "The Beer Can Theory of Creativity," 2000, http://cogprints.org/4768/, and L. Gabora, "Conceptual Closure: Weaving Memories into an Interconnected Worldview," in *Closure: Emergent Organizations and Their Dynamics*, ed. G. Van de Vijver and J. Chandler, *Annals of the New York Academy of Sciences* 901 (2000): 42–53.

BIBLIOGRAPHY

Abrams, M. D. "Genotypic and Phenotypic Variation as Stress Adaptations in Temperate Tree Species: A Review of Several Case Studies," *Tree Physiology* 14, no. 7 (1994): 833–42.

Albertson, R. C., J. T. Streelman, and T. D. Kocher. "Genetic Basis of Adaptive Shape Differences in the Cichlid Head." *Journal of Heredity* 94, no. 4 (2003): 291–301.

Argawal, A. A. "Phenotypic Plasticity in the Interactions and Evolution of Species." *Science* 294 (2001): 321–26.

Arnold, S. J. "Constraints on Phenotypic Evolution." *American Naturalist* 140, Supplement 1 (1992): S85–S107.

Ashburner, M. "Speculations on the Subject of Alcohol Dehydrogenase and Its Properties in *Drosophila* and Other Flies." *BioEssays* 20 (1998): 949–54.

Aspinwall, N. "Genetic Analysis of North American Populations of the Pink Salmon, *Oncorhyncus gorbuscha*, Possible Evidence for the Neutral Mutation-Random Drift Hypothesis." *Evolution* 28 (1974): 295–305.

Ayala, F. J. *Population and Evolutionary Genetics: A Primer*. Menlo Park, CA: Benjamin Cummings, 1982.

Ayala, F., and M. Coluzzi. "Chromosome Speciation: Humans, *Drosophila*, and Mosquitos." In *Systematics and the Origin of Species: On Ernst Mayr's 100th Anniversary*, edited by J. Hey, W. M. Fitch, and F. Ayala, 46–68. Washington, DC: National Academies Press, 2005.

Babcock, R. C., C. N. Mundy, and D. Whitehead. "Sperm Diffusion Models and In Situ Confirmation of Long-Distance Fertilization in the Free-Spawning Asteroid *Acanthaster planci*." *Biological Bulletin* 186, no. 1 (1994): 17–28.

BIBLIOGRAPHY

Baltosser, W. H. "Nectar Availability and Habitat Selection by Hummingbirds in Guadalupe Canyon." *Wilson Bulletin* 101, no. 4 (1989): 559–78.

Baltzer, F. Theodor Boveri. *The Life of a Great Biologist*. Translated from the German by Dorothea Rudnick. Berkeley: University of California Press, 1967.

Bakker, E. *An Island Called California*. Berkeley: University of California Press, 1984.

Bapteste, E., and Y. Boucher. "Lateral Gene Transfer Challenges Principles of Microbial Systematics." *Trends in Microbiology* 16, no. 5 (2008): 200–207.

Bapteste, E., Y. Boucher, J. Leigh, and W. F. Doolittle. "Phylogenetic Reconstruction and Lateral Gene Transfer." *Trends in Microbiology* 12, no. 9 (2004): 406–11.

Barluenga, M., K. N. Stölting, W. Salzburger, M. Muschick, and A. Meyer. "Sympatric Speciations in Nicaraguan Crater Lake Cichlid Fish." *Nature* 439 (2006): 719–23.

Barraclough, T. G., and S. Nee. "Phylogenetics and Speciation." *TRENDS in Ecology and Evolution* 16, no. 7 (2001): 391–99.

Beebe, W. "Notes on the Hercules Beetle *Dynastes Hercules* (Linn.), at Rancho Grande, Venezuela, with Special Reference to Combat Behavior." *Zoologica* 32 (1947): 109–16.

Bersaglieri, T., P. C. Sabeti, N. Patterson, T. Vanderploeg, S. F. Schaffner, J. A. Drake, M. Rhodes, D. E. Reich, and J. N. Hirschhorn. "Genetic Signatures of Strong Recent Positive Selection at the Lactase Gene." *American Journal of Human Genetics* 74, no. 6 (2004): 111–20.

Blows, M. W., and A. A. Hoffman. "A Reassessment of Genetic Limits to Evolutionary Change." *Ecology* 86, no. 6 (2005): 1371–84.

Boughman, J. W. "Divergent Sexual Selection Enhances Reproductive Isolation in Sticklebacks." *Nature* 411 (2001): 944–48.

Bowler, P. J. *Evolution: The History of an Idea*. 4th ed. Berkeley: University of California Press, 1989.

Bromham, L., and D. Penny. "The Modern Molecular Clock." *Nature Reviews Genetics* 4, no. 3 (2003): 216–24.

Bibliography

Brothwell, D., and P. Brothwell. *Food in Antiquity: A Survey of the Diet of Early Peoples*. Baltimore: Johns Hopkins University Press, 1969.

Brown, J. R. "Ancient Horizontal Gene Transfer." *Nature Reviews Genetics* 4 (2003): 121–32.

Bruni, L. E. "Gregory Bateson's Relevance to Current Molecular Biology." In *A Legacy for Living Systems: Gregory Bateson as Precursor to Biosemiotics*. Vol. 2, edited by J. P. Hoffmeyer, 93–119. New York: Springer, 2008.

Byrne, K., and R. A. Nichols. "*Culex pipiens* in London Underground Tunnels: Differentiation between Surface and Subterranean Populations." *Heredity* 82 (1999): 7–15.

Cahill, T. *How the Irish Saved Civilization*. Hinges of History Series, vol. 1. New York: Anchor Books, 1986.

Carroll, S. B. "Endless Forms: The Evolution of Gene Regulation and Morphological Diversity." *Cell* 101 (2000): 577–80.

———. *Endless Forms Most Beautiful: The New Science of Evo-Devo and the Making of the Animal Kingdom*. New York: W. W. Norton, 2005.

———. "Evo-Devo and an Expanding Evolutionary Synthesis: A Genetic Theory of Morphological Evolution." *Cell* 134 (2005): 25–36.

Carson, H. L., and A. R. Templeton. "Genetic Revolutions in Relation to Speciation Phenomena: The Founding of New Populations." *Annual Review of Ecology and Systematics* 15 (1984): 97–131.

Chai, P., and D. Millard. "Flight and Size Constraints: Hovering Performance of Large Hummingbirds under Maximal Loading." *Journal of Experimental Biology* 200 (1997): 2757–63.

Cheney, D. L., and R. M. Seyfarth. "Social and Non-Social Knowledge in Vervet Monkeys." *Philosophical Transactions of the Royal Society B* 308 (1985): 187–201.

Clutton-Brock, T. "Sexual Selection in Females." *Animal Behavior* 77, no. 1 (2009): 3–11.

———. "Sexual Selection in Males and Females." *Science* 318 (2007): 1882–85.

Costerton, J. W. "Cystic Fibrosis Pathogenesis and the Role of Biofilms in Persistent Infection." *Trends in Microbiology* 9, no. 2 (2001): 50–52.

BIBLIOGRAPHY

Coyne, J. A., and H. A. Orr. *Speciation*. Sunderland, MA: Sinauer, 2004.

Crespi, B. J. "The Evolution of Social Behavior in Microorganisms." *Trends in Ecology and Evolution* 16, no. 4 (2001): 178–83.

———. "Species Selection." In *Encyclopedia of the Life Sciences*. Vol. 17. London: Nature Publishing Group, 2002.

Crick, F. *What Mad Pursuit*. New York: Basic Books, 1990.

Crick, F. H. C., and J. D. Watson. "The Complementary Structure of Deoxyribonucleic Acid." *Proceedings of the Royal Society of London* A 223, no. 1152 (1954): 80–96.

Dalrymple, G. B. *The Age of the Earth*. Stanford, CA: Stanford University Press, 1991.

Dawkins, R. *The Extended Phenotype: The Gene as the Unit of Selection*. San Francisco: W. H. Freeman, 1982.

———. *The Selfish Gene*. Oxford: Oxford University Press, 1976.

———. "Universal Darwinism." In *Evolution: From Molecules to Men*, edited by D. S. Bendall, 403–25. Cambridge: Cambridge University Press, 1983.

Dean, D. R. "The Age of the Earth Controversy." *Annals of Science* 38 (1981): 435–56.

de Queiroz, K. "Ernst Mayr and the Modern Concept of Species." *Proceedings of the National Academy of Sciences of the USA* 102. Supplement 1 (2005): 6600–6607.

———. "Ernst Mayr and the Modern Concept of Species." In *Systematics and the Origin of Species: On Ernst Mayr's 100th Anniversary*, edited by J. Hey, W. M. Fitch, and F. Ayala, 243–66. Washington, DC: National Academies Press, 2005.

DeSalle, R., J. Gatesy, W. Wheeler, and D. Grimaldi. "DNA Sequences from a Fossil Termite in Oligo-Miocene Amber and Their Phylogenetic Implications." *Science* 257 (1992): 1933–36.

DeVantier, L. M. "Rafting of Tropical Marine Organisms on Buoyant Coralla." *Marine Ecology Progress*. Series 86 (1992): 301–2.

Dewitt, T. J., and S. M. Scheiner. *Phenotypic Plasticity: Functional and Conceptual Approaches*. Oxford: Oxford University Press, 2004.

Dodd, K. C. *North American Box Turtles: A Natural History*. Norman: University of Oklahoma Press, 2002.

Donald, M. *Origins of the Modern Mind: Three Stages in the Evolution of Culture and Cognition*. Cambridge, MA: Harvard University Press, 1991.

Doolan, S. P., and D. W. MacDonald. "Breeding and Juvenile Survival among Slender-Tailed Meerkats (*Suricata suricatta*) in the South-Western Kalahari: Ecological and Social Influences." *Journal of Zoology* 242 (1997): 309–27.

Douglas, A. E. *The Symbiotic Habit*. Princeton, NJ: Princeton University Press, 2010.

Edelman, G. *Bright Air, Brilliant Fire: On the Matter of the Mind*. New York: Basic Books, 1983.

Eldredge, N. "Cretaceous Meteor Showers, the Human Ecological "Niche," and the Sixth Extinction." In *Extinctions in Near Time: Causes, Contexts, and Consequences*, edited by R. D. E. MacPhee, 1–14. New York: Kluwer Academic/Plenum, 1999.

———. "Evolutionary Tempos and Modes: A Palaeontological Perspective." In *What Darwin Began: Modern Darwinian and Non-Darwinian Perspectives on Evolution*, edited by L. R. Godfrey, 113–17. Boston: Allyn & Bacon, 1985.

Elmer, K. R., C. Reggio, T. Wirth, E. Verheyen, W. Salzburger, and A. Meyer. "Pleistocene Desiccation in East Africa Bottlenecked but Did Not Extirpate the Adaptive Radiation of Lake Victoria Haplochromine Cichlid Fishes." *Proceedings of the National Academy of Sciences* (USA) 106, no. 32 (2009): 13404–9.

Endler, J. A. *Natural Selection in the Wild*. Princeton, NJ: Princeton University Press, 1986.

Fernald, R. D. "Casting a Genetic Light on the Evolution of Eyes." *Science* 313 (2006): 1914–18.

Feulner, P. G. D., M. Plath, J. Engelmann, F. Kirschbaum, and R. Tiedemann. "Electrifying Love: Electric Fish Use Species-Specific Discharge for Mate Recognition." *Biology Letters* 5, no. 2 (2009): 225–28.

BIBLIOGRAPHY

Frankham, R. "Do Island Populations Have Less Genetic Variation Than Mainland Populations?" *Heredity* 78 (1997): 311–27.

Frazer, N. B. "Sea Turtle Conservation and Halfway Technology." *Conservation Biology* 6, no. 2 (2003): 179–84.

Friedberg, E. C. "Mutation as a Phenotype." In *The Implicit Genome*, edited by L. H. Caporale, 39–56. Oxford: Oxford University Press, 2006.

Friedberg, E. C., G. C. Walker, and W. Seide. *DNA Repair and Mutagenesis*. Washington, DC: ASM, 1995.

Futuyama, D. J., and G. C. Mayer. "Non-Allopatric Speciation in Animals." *Systematic Zoology* 29, no. 3 (1980): 254–71.

Gabora, L. 2000. "Conceptual Closure: Weaving Memories into an Interconnected Worldview." In *Closure: Emergent Organizations and Their Dynamics*, edited by G. Van de Vijver and J. Chandler, 42–53. *Annals of the New York Academy of Sciences* 901 (2000).

Gal, R., and F. Libersat. "A Parasitoid Wasp Manipulates the Drive for Walking of Its Cockroach Prey." *Current Biology* 18 (2008): 877–82.

Galton, F. "Experiments in Pangenesis, by Breeding from Rabbits of a Pure Variety, Into Whose Circulation Blood Taken from Other Varieties Had Previously Been Transfused." *Proceedings of the Royal Society* XIX (1871): 393–410.

Geison, G. L. "Darwin and Heredity: The Evolution of His Hypothesis of Pangenesis." *Journal of the History of Medicine and Allied Sciences* 24, no. 4 (1969): 365–411.

Gershwin, L.-A. "Clonal and Population Variation in Jellyfish Symmetry." *Journal of the Marine Biological Association of the United Kingdom* 79 (1999): 993–1000.

Gogarten, J. P., W. F. Doolitle, and J. G. Lawrence. "Prokaryotic Evolution in Light of Gene Transfer." *Molecular Biological Evolution* 19, no. 12 (2002): 2226–38.

Gogarten, J. P., and F. Townsend. "Horizontal Gene Transfer, Genome Innovation, and Evolution." *Nature Reviews Microbiology* 3 (2005): 679–87.

Gogarten, M. B., J. P. Gogarten, and L. Oldenzewski. *Horizontal Gene Transfer: Genomes in Flux*. New York: Humana Press, 2009.

Goldenfeld, N., and C. Woese. "Biology's Next Revolution." *Nature* 445 (2007): 369.

Gould, S. J. "The Disparity of the Burgess Shale Arthropod Fauna and the Limits of Cladistic Analysis." *Paleobiology* 17 (1991): 411–23.

Gould, S. J., and N. Eldredge, eds. *Genetics and the Origin of Species by Theodosius Dobzhansky*. Columbia Classics in Evolution Series. New York: Columbia University Press, 1982.

Gould, S. J., and R. C. Lewontin. "The Spandrels of San Marco and the Panglossian Paradigm: A Critique of the Adaptationist Programme." *Proceedings of the Royal Society of London* B 205 (1979): 581–98.

Grant, P. R., B. R. Grant, and K. Petren. "The Allopatric Phase of Speciation: The Sharpbeaked Ground Finch (*Geospiza difficilis*) on the Galapagos Islands." *Biological Journal of the Linnean Society* 69 (2000): 287–317.

Grantham, T. A. "Hierarchical Approaches to Macroevolution: Recent Work on Species Selection and the 'Effect Hypothesis.'" *Annual Review of Ecology and Systematics* 26 (1995): 301–21.

Green, R. E., J. Krause, A. W. Briggs, T. Maricic, U. Stenzel, M. Kircher, N. Patterson, H. Li, W. Zhai, M. Hsi-Yang Fritz, N. F. Hansen, E. Y. Durand, A.-S. Malaspinas, J. D. Jensen, T. Marques-Bonet, C. Alkan, K. Prüfer, M. Meyer, H. A. Burbano, J. M. Good, R. Schultz, A. Aximu-Petri, A. Butthof, B. Höber, B. Höffner, M. Siegemund, A. Weihmann, C. Nusbaum, E. S. Lander, C. Russ, N. Novod, J. Affourtit, M. Egholm, C. Verna, L. Rudan, D. Brajkovic, Z. Kucan, I. Gusic, V. B. Doronichev, L. V. Golovanova, C. Lalueza-Fox, M. de la Rasilla, J. Fortea, A. Rosas, R. W. Schmitz, P. L. F. Johnson, E. E. Eichler, D. Falush, E. Birney, J. C. Mullikin, M. Slatkin, R. Nielsen, J. Kelso, M. Lachmann, D. Reich, and S. Pääbo. "A Draft Sequence of the Neandertal Genome." *Science* 328, no. 5979 (2010): 710–22.

Green, R. E., J. Krause, S. E. Ptak, A. W. Briggs, M. T. Ronan, J. F. Simons, L. Du, M. Egholm, J. M. Rothberg, M. Paunovic, and S. Pääbo. "Analysis of One Million Base Pairs of Neanderthal DNA." *Nature* 444 (2006): 330–36.

Grell, E. H., K. B. Jacobson, and J. B. Murphy. "Alterations of Genetic Material for Analysis of Alcohol Dehydrogenase Isozymes of *Drosophila melanogaster*." *Annals of the New York Academy of Sciences* 151 (1968): 441–45.

Gross, M. R. "Disruptive Selection for Alternative Life Histories in Salmon." *Nature* 313 (1985): 47–48.

Guttman, B., A. Griffiths, D. Suzuki, and T. Cullis. *Genetics: A Beginner's Guide*. Oxford: Oneworld, 1982.

Hanlon, R. T. "Mating Systems and Sexual Selection in the Squid *Loligo*: How Might Commercial Fishing on Spawning Squids Affect Them?" *CalCOFL Report* 39 (1998): 92–100.

Hansen, T. F., and D. Houle. "Evolvability, Stabilizing Selection, and the Problem of Stasis." In *Phenotypic Integration: Studying Ecology and the Evolution of Complex Phenotypes*, edited by M. Pigliucci and K. Preston, 130–50. Oxford: Oxford University Press, 2004.

Harshey, R. M. "Bacterial Motility on a Surface: Many Ways to a Common Goal." *Annual Review of Microbiology* 57 (2003): 249–73.

Hart, D., and R. W. Sussman. *Man the Hunted: Primates, Predators, and Human Evolution*. Boulder, CO: Westview Press, 2005.

Hatle, J. D., D. W. Borst, and S. A. Juliano. "Plasticity and Canalization in the Control of Reproduction in the Lubber Grasshopper." *Integrative Comparative Biology* 43 (2003): 635–45.

Heard, E., S. Tishkoff, J. A. Todd, M. Vidal, G. P. Wagner, J. Want, D. Weigel, and R. Young. "Ten Years of Genetics and Genomics: What Have We Achieved and Where Are We Heading?" *Nature Reviews Genetics* (2010) advance online publication. doi:10.1038/nrg2878. Accessed March 10, 2011.

Helmuth, B., R. R. Veit, and R. Holberton, "Long-Distance Dispersal of a Subantarctic Brooding Bivalve (*Gaimardia trapesina*) by Kelp-Rafting." *Marine Biology* 120 (1994): 421–26.

Henshilwood, C. S., F. d'Errioco, R. Yates, Z. Jacobs, C. Tribolo, G. A. T. Duller, N. Mercier, J. C. Sealy, H. Valladas, I. Watts, and A.G. Wintle. "Emergence of Modern Human Behavior: Middle Stone Age Engravings from South Africa." *Science* 295, no. 5558 (2002): 1278–80.

Bibliography

Herodotus, *The Histories*, Book II in *Herodotus*. English translation by A. D. Godley. Cambridge, MA: Harvard University Press, 1920.

Hey, J., W. M. Fitch, and F. Ayala, eds. *Systematics and the Origin of Species on Ernst Mayr's 100th Anniversary*. Washington, DC: National Academies Press, 2005.

Hodge, S. J., A. Manica, T. P. Flower, and T. H. Clutton-Brock. "Determinants of Reproductive Success in Dominant Female Meerkats." *Journal of Animal Ecology* 77 (2008): 92–102.

Holmes, R. *The Age of Wonder: How the Romantic Generation Discovered the Beauty and Terror of Science*. New York: Vintage Books, 2009.

Hori, M. "Frequency-Dependent Natural Selection in the Handedness of Scale-Eating Cichlid Fish." *Science* 260 (1993): 216–19.

Howard, J. C. "Why Didn't Darwin Discover Mendel's Laws?" *Journal of Biology* 8 (2008): 15.1–15.8.

Hull, D. "Individuality and Selection." *Annual Review of Ecology and Systematics* 11 (1980): 311–32.

Hurst, L. D. "Genetics and the Understanding of Selection." *Nature Reviews Genetics* 10 (2009): 83–93.

Huxley, J. *Evolution: The Modern Synthesis*. London: Allen & Unwin, 1942.

Huxley, T. H. *Darwiniana: Essays by Thomas Henry Huxley*. Vol. 2. London: Macmillan, 1863.

Irwin, D. E. "Song Variation in an Avian Ring Species." *Evolution* 54, no. 3 (2000): 998–1010.

———. "Speciation by Distance in a Ring Species." *Science* 307 (2005): 414–16.

Joikel, P., and F. J. Martinelli. "The Vortex Model of Coral Reef Biogeography." *Journal of Biogeography* 19 (1992): 449–58.

Jones, G., and E. C. Teeling. "The Evolution of Echolocation in Bats." *Trends in Ecology and Evolution* 21, no. 3 (2006): 149–56.

Keeling, P. J., and J. D. Palmer. "Horizontal Gene Transfer in Eukaryotic Evolution." *Nature Reviews Genetics* 9 (2008): 605–18.

Kehoe, A. "Modern Antievolutionism: The Scientific Creationists." In *What*

Darwin Began: Modern Darwinian and Non-Darwinian Perspectives on Evolution, edited by L. R. Godfrey, 156–85. Boston: Allyn & Bacon, 1985.

Kerr, R. A. "Evolution: New Mammal Data Challenge Evolutionary Pulse Theory." *Science* 273, no. 5274 (1996): 431–32.

Kilias, G., S. N. Alahiotis, and M. Pelecanos. "A Multifactorial Genetic Investigation of Speciation Theory Using *Drosophila melanogaster*." *Evolution* 34, no. 4 (1980): 730–37.

Knowlton, N., J. L. Maté, H. M. Guzmán, R. Rowan, and J. Jara. "Direct Evidence for Reproductive Isolation among the Three Species of the *Montastraea annularis* Complex in Central America (Panamá and Honduras)." *Marine Biology* 127, no. 4 (1997): 705–11.

Knowlton, N., L. A. Weigt, L. A. Solórzano, D. K. Mills, and E. Bermingham. "Divergence in Proteins, Mitochondrial DNA, and Reproductive Compatibility across the Isthmus of Panama." *Science* 260 (1993): 1629–32.

Kocher, T. D. "Adaptive Evolution and Explosive Speciation: The Cichlid Fish Model." *Nature Reviews Genetics* 5 (2004): 288–98.

Koonin, E. V. "Darwinian Evolution in the Light of Genomics." *Nucleic Acids Research* 37, no. 4 (2009): 1011–34.

———. "The Origin at 150: Is a New Evolutionary Synthesis in Sight?" *Trends in Genetics* 25, no. 11 (2009): 473–75.

Koonin, E. V., and Y. Boucher. "Is Evolution Darwinian or/and Lamarckian?" *Biology Direct* 4 (2009). doi:10.1186/1745-6150-4-42. Accessed March 10, 2011.

Kovacs, E. H., and I. S. Sas. "Cannibalistic Behaviour of *Epidalea* (Bufo) *viridis* Tadpoles in an Urban Breeding Habitat." *North-Western Journal of Zoology* 5, no. 1 (2009): 206–8.

Langerhans, R. B., M. E. Gifford, and E. O. Joseph. "Ecological Speciation in Gambusia Fishes." *Evolution* 61, no. 9 (2007): 2056–74.

Larsen, T. "Polar Bear Denning and Cub Production in Svalbard, Norway." *Journal of Wildlife Management* 49, no. 2 (1985): 320–26.

Latham, R. E., trans. and ed. *Lucretius: On the Nature of the Universe*. London: Penguin Books, 1982.

Leakey, R., and R. Lewin. *The Sixth Extinction: Biodiversity and Its Survival.* London: Weidenfeld & Nicolson, 1996.

Lessios, H. A. "The Great American Schism: Divergence of Marine Organisms after the Rise of the Central American Isthmus." *Annual Review of Ecology, Evolution, and Systematics* 39 (2008): 63–91.

Levins, R. *Evolution in Changing Environments.* Princeton, NJ: Princeton University Press, 1968.

Lewin, R., and R. Foley. *Principles of Human Evolution.* London: Blackwell Science, 2004.

Li, Y., Z. Liu, P. Shi, and J. Zhang. "The Hearing Gene Prestin Unites Echolocating Bats and Whales." *Current Biology* 20, no. 2 (2008): R55–R56.

Logan, G. A., J. J. Boon, and G. Eglinton. "Structural Biopolymer Preservation in Miocene Leaf Fossils from the Clarkia Site, Northern Idaho." *Proceedings of the National Academy of Sciences USA* 90 (1993): 2246–50.

Lopez, B. "The Naturalist." *Orion* magazine. Autumn 2001. http://www.orionmagazine.org/index.php/articles/article/91. Accessed March 10, 2011.

Lovejoy, A. O. *The Great Chain of Being: A Study of the History of an Idea.* Cambridge, MA: Harvard University Press, 1936.

Maclean, I. *The Renaissance Notion of Women: A Study in the Fortunes of Scholasticism and Medical Science in European Intellectual Life.* New York: Cambridge University Press, 1980.

Macnab, R. "How Bacteria Assemble Flagella." *Annual Review of Microbiology* 57 (2003): 77–100.

MacPhee, R., and C. Flemming. "Requiem Aeternum: The Last Five Hundred Years of Mammalian Species Extinctions." In *Extinctions in Near Time: Causes, Contexts, and Consequences,* edited by R. D. E. MacPhee, 333–72. New York: Kluwer Academic/Plenum, 1999.

Madsen, P. T., M. Johnson, N. Aguilar de Soto, W. M. X. Zimmer, and P. Tyack. "Bisonar Performance of Foraging Beaked Whales (*Mesoplodon densirostris*)." *Journal of Experimental Biology* 208 (2005): 181–94.

Madsen, T., M. Olsson, H. Wittzell, B. Stille, A. Gullberg, R. Shine, S. Andersson, and H. Tegelström. "Population and Genetic Diversity in Sand

Lizards (*Lacerta agilis*) and Adders (*Vipera berus*)." *Biological Conservation* 94 (2000): 257–62.

Malherbe, Y., A. Kamping, W. van Delden, and L. van de Zande. "ADH Enzyme Activity and Adh Gene Expression in *Drosophila melanogaster* Lines Differentially Selected for Increased Alcohol Tolerance." *Journal of Evolutionary Biology* 18 (2005): 811–19.

Mallet, J. "A Species Definition for the Modern Synthesis." *Trends in Ecology and Evolution* 10 (1995): 294–99.

Margulis, L., and D. Sagan. *Acquiring Genomes: A Theory of the Origins of Species*. New York: Basic Books, 2002.

Marshall, C. R. "Explaining the Cambrian 'Explosion' of Animals." *Annual Review of Earth and Planetary Sciences* 34 (2006): 355–84.

May, R. M. "How Many Species?" *Philosophical Transactions of the Royal Society of London* B 330 (1990): 293–304.

Mayr, E. "What Is a Species and What Is Not?" *Philosophy of Science* 63 (1996): 262–77.

———. *What Makes Biology Unique? Considerations on the Autonomy of a Scientific Discipline*. Cambridge: Cambridge University Press, 2004.

Mayr, E., and Provine, W. B., eds. *The Evolutionary Synthesis: Perspectives on the Unification of Biology*. With a new preface by Ernst Mayr. Cambridge, MA: Harvard University Press, 1998.

McDonald, M. A. "Early African Pastoralism: View from Dakhleh Oasis (South Central Egypt)." *Journal of Anthropological Archaeology* 17, no. 2 (1998): 124–42.

McGraw, W. S., C. Cooke, and S. Shultz. "Primate Remains from African Crowned Eagle (*Stephanoaetus coronatus*) Nests in Ivory Coast's Tai Forest: Implications for Primate Predation and Early Hominid Taphonomy in South Africa." *American Journal of Physical Anthropology* 131 (2006): 151–65.

McKinnon, J. S., and H. D. Rundle. "Speciation in Nature: The Threespine Stickleback Model System." *TRENDS in Ecology & Evolution* 17, no. 110 (2002): 480–88.

Melendez-Ackerman, E., D. R. Campbell, and N. M. Waser. "Hummingbird Behavior and Mechanisms of Selection on Flower Color in Ipomopsis." *Ecology* 78, no. 8 (1997): 2532–41.

Menotti-Raymond, M., and S. J. O'Brien. "Dating the Genetic Bottleneck of the African Cheetah." *Proceedings of the National Academy of Sciences* 90, no. 8 (1993): 3172–76.

Michod, R. E. *Darwinian Dynamics: Evolutionary Transitions in Fitness and Individuality*. Princeton, NJ: Princeton University Press, 1999.

Milton, J. *The Annotated Milton: Complete English Poems*, edited by B. Raffel. New York: Bantam, 1999.

Minkoff, E. C. *Evolutionary Biology*. Reading, MA: Addison-Wesley, 1983.

Mithen, S. *The Prehistory of the Mind*. London: Thames and Hudson, 1999.

Myers, G., I. Paulsen, and C. Fraser. "The Role of Mobile DNA in the Evolution of Prokaryotic Genomes." In *The Implicit Genome*, edited by L. H. Caporale, 121–37. New York: Oxford University Press, 2006.

Nagata, N., K. Kubota, Y. Takami, and T. Sota. "Historical Divergence of Mechanical Isolation Agents in the Ground Beetle *Carabus arrowianus* as Revealed by Phylogeographical Analyses." *Molecular Ecology* 18, no. 7 (2009): 1408–21.

Nettle, D. "The Evolution of Personality Variation in Human and Other Animals." *American Psychologist* 61, no. 6 (2006): 622–31.

Niemiller, M. J., B. J. Fitzpatrick, and B. T. Miller. "Recent Divergence with Gene Flow in Tennessee Cave Salamanders (*Plethodontidae: Gyrinophilus*) Inferred from Gene Genealogies." *Molecular Ecology* 17 (2008): 2258–75.

Numbers, R. L. *Darwinism Comes to America*. Cambridge, MA: Harvard University Press, 1998.

O'Brien, S. J., M. E. Roelke, L. Marker, A. Newman, C. A. Winkler, D. Meltzer, L. Colly, J. F. Evermann, M. Bush, and D. E. Wildt. "Genetic Basis for Species Vulnerability in the Cheetah." *Science* 227, no. 4693 (1985): 1428–34.

Oehser, P. H. "Louis Jean Pierre Vieillot (1748–1831)." *Auk* 65 (1948): 568–76.

Olby, R. *The Path to the Double Helix*. London: Dover, 1984.

Olsen, G. J., and C. Woese. "Archaeal Genomics: An Overview." *Cell* 89 (1997): 991–94.

Olson, E. N. "Gene Regulatory Networks in the Evolution and Development of the Heart." *Science* 313 (2006): 1922–27.

Pääbo, S. "Ancient DNA." In *DNA: Changing Science and Society*, edited by T. Krude, 68–87. Cambridge: Cambridge University Press, 2004.

Pagel, M. "Natural Selection 150 Years On." *Nature* 457 (2009): 808–11.

Parjeko, K. "Embryology of *Chaoborus*-Induced Spines in *Daphnia pulex*." *Hydrobiologia* 231 (1992): 77–84.

Paul, J. H. "Microbial Gene Transfer: An Ecological Perspective." *Journal of Molecular Biotechnology* 1, no. 1 (1999): 45–50.

Paweletz, N. "Walther Flemming: Pioneer of Mitosis Research." *Nature Reviews Molecular Cell Biology* 2 (2001): 72–75.

Pennisi, E. "Changing a Fish's Bony Armor in the Wink of a Gene." *Science* 304 (2004): 1736–39.

Pierce, N. E., and P. S. Mead. "Parasitoids as Selective Agents in the Symbiosis between Lycaenid Butterfly Larvae and Ants." *Science* 211 (1981): 1185–87.

Pinto-Correia, C. *The Ovary and Eve: Egg and Sperm and Preformation*. Chicago: University of Chicago Press, 1998.

Porta, M. "The Genome Sequence Is a Jazz Score." *International Journal of Epidemiology* 32, no. 1 (2003): 29–31.

Porter, M. L., and K. A. Crandall. "Lost Along the Way: The Significance of Evolution in Reverse." *TRENDS in Ecology and Evolution* 18, no. 10 (2003): 541–47.

Portier, C., M. Festa-Bianchet, J.-M. Gaillard, J. T. Jorgenson, and N. G. Yoccoz. "Effects of Density and Weather on Survival of Bighorn Sheep Lambs (*Ovis canadensis*)." *Journal of Zoology* 245 (1998): 271–78.

Ramsey, J., H. D. Bradshaw Jr., and Schemske, D. W. "Components of Reproductive Isolation between the Monkeyflowers *Mimulus lewisii* and *M. cardinali* (*Phrymaceae*)." *Evolution* 57 (2003): 1520–34.

Reuffler, C., T. J. Van Dooren, O. Leimar, and P. A. Abrams. "Disruptive Selection and Then What?" *Trends in Ecology and Evolution* 21, no. 5 (2006): 238–45.

Reznick, D. A., H. Bryga, and J. A. Endler. "Experimentally Induced Life-History Evolution in a Natural Population." *Nature* 346 (1990): 357–59.

Ridley, M. *Genome: The Autobiography of a Species in 23 Chapters*. New York: Harper Perennial, 2006.

Rose, M. R., and T. H. Oakley. "The New Biology: Beyond the Modern Synthesis." *Biology Direct* 2 (2007). doi:10.1186/1745-6150-2-30. Accessed March 10, 2011.

Roux, J., and M. Robinson-Rechavi. "Developmental Constraints on Vertebrate Genome Evolution." *PLoS Genet* 4 (12): e1000311.doi:10.1371/journal.pgen.1000311. Accessed March 10, 2011.

Ruiz-Mirazo, K., J. Pereto, and A. Moreno. "A Universal Definition of Life: Autonomy and Open-Ended Evolution." *Origins of Life and Evolution of the Biosphere* 34 (2004): 323–46.

Rundle, H. D., L. H. Nagel, J. W. Boughman, and D. Schluter. "Natural Selection and Parallel Sympatric Speciation in Sticklebacks." *Science* 287 (2000): 306–7.

Sampson, J. A., and E. S. C. Weiner, eds. *Oxford English Dictionary*. Oxford: Clarendon Press, 1989.

Sato, A., H. Tichy, C. O'hUigin, P. R. Grant, B. R. Grant, and J. Klein. "On the Origin of Darwin's Finches." *Molecular Biology and Evolution* 18, no. 3 (2000): 299–311.

Savage, J. M. *Evolution*. 3rd ed. New York: Holt, Rinehart and Winston, 1977.

Scott, E. *Evolution and Creationism*. Berkeley: University of California Press, 2008.

Schemske, D. W., and H. D. Bradshaw Jr. "Pollinator Preference and the Evolution of Floral Traits in Monkeyflowers (*Mimulus*)." *Proceedings of the National Academy of Sciences* (USA) 96 (2003): 11910–15.

Schilthuizen, M. *Frogs, Flies, and Dandelions—Speciation, The Origin of New Species*. Oxford: Oxford University Press, 2001.

Seehausen, O., J. J. M. van Alphen, and F. White. "Cichlid Fish Diversity Threatened by Eutrophication That Curbs Sexual Selection." *Science* 277 (1997): 1811.

Shapiro, M. D., M. E. Marks, C. L. Peichel, B. J. Blackman, K. S. Nereng, B. Jonsson, D. Schluter, and D. M. Kingsley. "Genetic and Developmental Basis of Evolutionary Pelvic Reduction in Threespine Sticklebacks." *Nature* 428 (2004): 717–23.

Shavit, A., and R. L. Millstein. "Group Selection Is Dead! Long Live Group Selection!" *BioScience* 58, no. 7 (2008): 574–75.

Shen, Y.-Y., J. Liu, D. M. Irwin, and Y.-P. Zhang. "Parallel and Convergent Evolution of the Dim-Light Vision Gene *RH1* in Bats (Order: Chiroptera)." *PLoS Biology* 15, no. 1 (2010). doi:10.1371/journal.pone.0008838. Accessed March 10, 2011.

Sherratt, A. "Cups That Cheered: The Introduction of Alcohol into Prehistoric Europe." In *Economy and Society in Prehistoric Europe: Changing Perspectives,* edited by A. Sherratt, 376–402. Princeton, NJ: Princeton University Press, 1997.

Silverstein, J. T., and W. K. Hershberger. "Genetic Parameters of Size Pre- and Post-Smoltification in Coho Salmon (*Oncorhynchus kisutch*)." *Aquaculture* 128, no. 1 (1994): 67–77.

Simpson, S. J., E. Despland, B. F. Hagele, and T. Dodgson. "Gregarious Behavior in Desert Locusts Is Evoked by Touching Their Back Legs." *Proceedings of the National Academy of Sciences* (USA) 98, no. 7 (2001): 3895–97.

Smith, C. M. "Rise of the Modern Mind." *Scientific American MIND* (August 2006): 73–79.

Smith, C. M., and C. Sullivan. *The Top Ten Myths about Evolution*. Amherst, NY: Prometheus Books, 2006.

Smith, J. M., N. H. Smith, M. O'Rourke, and B. G. Spratt. "How Clonal Are Bacteria?" *Proceedings of the National Academy of Sciences of the United States of America* 90 (1993): 4384–88.

Smith, T. B. "Disruptive Selection and the Genetic Basis of Bill Size Polymorphism in the African Finch *Pyrenestes*." *Nature* 363 (1993): 618–20.

Sober, E. *Evidence and Evolution*. Cambridge: Cambridge University Press, 2008.

Sokolowski, M. B. "Genes for Normal Behavioral Variation: Recent Clues from Flies and Worms." *Neuron* 21 (1998): 463–66.

Sorhannus, U., E. J. Fenster, A. Hoffman, and L. Burckle. "Iterative Evolution in the Diatom Genus *Rhizosolenia* (Ehrenberg)." *Lethaia* 24, no. 1 (2001): 39–44.

Sota, T., and K. Kubota. "Genital Lock-and-Key as a Selective Agent against Hybridization." *Evolution* 52 (1998): 1507–13.

Southwood, T. R. E. "Interactions of Plants and Animals: Patterns and Processes." *Oikos* 44 (1985): 5–11.

Spight, T. "Availability and Use of Shells by Intertidal Hermit Crabs." *Biology Bulletin* 152 (1977): 120–33.

Stearns, S. C. "The Evolutionary Significance of Phenotypic Plasticity." *Bioscience* 37, no. 7 (1989): 436–45.

Suttle, C. A. "Viruses in the Sea." *Nature* 437 (2005): 356–61.

Sweitzer, R. A., and D. H. Van Vuren. *Rooting and Foraging Effects of Wild Pigs on Tree Regeneration and Acorn Survival in California's Oak Woodland Ecosystems*. USDA Forest Service General Technical Report PSW-GTR-184, 2002.

Sykes, B. "Using Genes to Map Population Structure and Origins." In *The Human Inheritance*, edited by B. Sykes, 93–117. Oxford: Oxford University Press, 1999.

Teeling, E. C. "Hear, Hear; The Convergent Evolution of Echolocation in Bats." *Trends in Ecology and Evolution* 24, no. 7 (2009): 351–54.

Telford, S. R., and P. I. Webb. "The Energetic Cost of Copulation in a Polygynandrous Millipede." *Journal of Experimental Biology* 201 (1998): 1847–49.

Templeton, A. R. "Out of Africa Again and Again." *Nature* 416 (2002): 45–51.

Thomas, G. L., and R. E. Thorne. "Night-Time Predation by Steller Sea Lions." *Nature* 411 (2001): 1013.

Thompson, J. N. "The Evolution of Species Interactions." *Science* 284 (1999): 2116–18.

Thompson, J. N., S. L. Nuismer, and R. Gomulkiewicz. "Coevolution and Maladaptation." *Integrative and Comparative Biology* 42 (2002): 381–87.

Tishkoff, S. A., F. A. Reed, A. Ranciaro, B. F. Voight, C. C. Babbitt, J. S. Silverman, K. Powell, H. M. Mortensen, J. B. Hirbo, M. Osman, M. Ibrahim, S. A. Omar, G. Lema, T. B. Nyambo, J. Ghori, S. Bumpstead, J. K. Pritchard, G. A. Wray, and P. Deloukas. "Convergent Adaptation of Human Lactase Persistence in Africa and Europe." *Nature Genetics* 39, no. 1 (2007): 31–40.

Tuchman, B. *A Distant Mirror: The Calamitous Fourteenth Century*. New York: Ballantine, 1987.

Tyndall, J. *Address Delivered before the British Association Assembled at Belfast, with Additions*. London: Longmans and Green, 1874.

Vermeij, G. J. *Biogeography and Adaptation*. Cambridge, MA: Harvard University Press, 1978.

———. "The Evolutionary Interaction among Species: Selection, Escalation, and Coevolution." *Annual Review of Ecology and Systematics* 25 (1994): 219–36.

Vrba, E. "Mammals as a Key to Evolutionary Theory." *Journal of Mammalogy* 73, no. 1 (1992): 1–28.

———. "Mass Turnover and Heterochrony Events in Response to Physical Change." *Paleobiology* 31 (2005):157–74.

Vrba, E. S. "On the Connections between Palaeoclimate and Evolution." In *Palaeoclimate and Evolution with Emphasis on Human Origins*, edited by E. S. Vrba, G. H. Denton, T. C. Partridge, and L. H. Burckle, 24–45. New Haven, CT: Yale University Press, 1995.

Warrick, D. R., B. W. Tobalske, and D. R. Powers. "Aerodynamics of the Hovering Hummingbird." *Nature* 435 (2005): 1094–97.

Waser, N., and M. V. Price. "Pollinator Choice and Stabilizing Selection for Flower Color in *Delphinium nelsonii*." *Evolution* 35, no. 2 (1983): 376–90.

Watson, J. D. *The Double Helix*. New York: Atheneum, 1968.

Watson, J. D., with A. Berry. *DNA: The Secret of Life*. New York: Alfred A. Knopf, 2003.

Webb, J. K., G. P. Brown, and R. Shine. "Body Size, Locomotor Speed, and Antipredator Behaviour in a Tropical Snake (*Tropidonophis mairii, Colubridae*): The Influence of Incubation Environments and Genetic Factors." *Functional Ecology* 15 (2001): 561–68.

Webb, W. C., W. J. Boarman, and T. Rotenberry. "Common Raven Juvenile Survival in a Human-Augmented Landscape." *Condor* 106 (2004): 517–28.

Weinberg, J. R., V. R. Starczak, and D. Jörg. "Evidence for Rapid Speciation Following a Founder Effect in the Laboratory." *Evolution* 46, no. 4 (1992): 1214–20.

Weiner, J. *The Beak of the Finch: A Story of Evolution in Our Time*. New York: Vintage/Random House, 1995.

Wells, D. J. "Muscle Performance in Hummingbirds." *Journal of Experimental Biology* 178 (1993): 39–57.

West, S. A., S. P. Diggle, A. Buckling, A. Gardner, and A. S. Griffin. "The Social Lives of Microbes." *Annual Review of Ecology, Evolution, and Systematics* 38 (2007): 53–77.

White, M. D. *Modes of Speciation*. San Francisco: W. H. Freeman, 1978.

Williams, G. C. *Natural Selection: Domains, Levels, and Challenges*. New York: Oxford University Press, 1992.

Wilson, E. O. "Introductory Essay: Systematics and the Future of Biology." In *Systematics and the Origin of Species: On Ernst Mayr's 100th Anniversary*, edited by J. Hey, W. M. Fitch, and F. Ayala, 1–8. Washington, DC: National Academies Press, 2005.

———. *Sociobiology: The New Synthesis*. Cambridge, MA: Belknap/Harvard University Press, 1975.

Witkin, E. S. "Ultraviolet Mutagenesis and the SOS Response in *Escherichia coli*: A Personal Perspective." *Environmental and Molecular Mutagenesis* 14. Supplement 16 (1989): 30–34.

Woese, C. "A New Biology for a New Century." *Microbiology and Molecular Biology Reviews* (June 2004): 173–86.

Wolf, J. B. W., C. Harrod, S. Brunner, S. Salazar, F. Trillmich, and D. Tautz. "Tracing Early Stages of Species Differentiation: Ecological, Morphological, and Genetic Divergence of Galapagos Sea Lion Populations." *BMC Evolutionary Biology* 8 (2008): 150. doi:10.1186/1471-2148-8-150. Accessed March 10, 2011.

Wood, T. E., and L. H. Rieseberg. "Speciation: Introduction." In *Encyclopedia of the Life Sciences*. Vol. 17. Nature Publishing Group, London, 2002, pp. 415–22.

Wyatt, T. D. *Pheromones and Animal Behavior: Communication by Smell and Taste*. Cambridge: Cambridge University Press, 2008.

Xu, L., H. Chen, X. Hu, R. Zhang, Z. Zhang, and Z. W. Luo. "Average Gene Length Is Highly Conserved in Prokaryotes and Eukaryotes and Diverges Only between the Two Kingdoms." *Molecular Biology and Evolution* 23, no. 6 (2006): 1107–8.

Young, A., T. Boyle, and T. Brown. "The Population Genetic Consequences of Habitat Fragmentation for Plants." *TREE* 10, no. 11 (1998): 413–18.

Zachar, I., and E. Szathmary. "A New Replicator: A Theoretical Framework for Analysing Replication." *BioMed Central Biology* 8, no. 21 (2010): 1–26.

Zheng, J., W. Shen, D. Z. Z. He, K. B. Long, L. D. Madison, and P. Dallos. "Prestin Is the Motor Protein of Cochlear Outer Hair Cells." *Nature* 405 (2000): 149–55.

Zillber-Rosenberg, I., and E. Rosenberg. "Role of Microorganisms in the Evolution of Animals and Plants: The Hologenome Theory of Evolution." *FEMS Microbiology Review* 32 (2008): 723–35.

Zimmer, C. *Evolution: The Triumph of an Idea*. New York: HarperCollins, 2001.

Zyll de Jong, C. G., C. Gates, H. Reynolds, and W. Olson. "Phenotypic Variation in Remnant Populations of North American Bison." *Journal of Mammalogy* 76, no. 2 (1995): 391–405.

WEBSITES FOR MORE ON EVOLUTION

In addition to the many books and articles I've listed in the notes section, there are some great websites with quality material that can help you learn more about the world of living things. Here are just a handful:

At the ***Encyclopedia of Life*** you can look up any life-form by entering either the common ("Nile crocodile") or the scientific ("*Crocodylus niloticus*") name. If the species is included in the database, you can see pictures of it and page through comments on its basic characteristics, ecology, reproduction, and genetics, as well as find out where to learn more about the species. The *Encyclopedia of Life* is the brainchild of evolutionary biologist E. O. Wilson. While not all Wikipedia entries are monitored for accuracy, at EOL (you can sign up to modify), "[e]xpert curators ensure quality of the core collection by authenticating materials submitted by diverse projects and individual contributors." http://www.eol.org.

The ***Tree of Life Web Project*** is another online encyclopedia. From the website: "The Tree of Life Web Project (*ToL*) is a collaborative effort of biologists and nature enthusiasts from around the world. On more than 10,000 World Wide Web pages, the project provides information about biodiversity, the characteristics of different groups of organisms, and their evolutionary history. Each page contains information about a particular group. ToL pages are linked one to another hierarchically, in the form of the evolutionary tree of life. Starting with 'Life on Earth' and moving out along diverging branches to individual species, the 'Structure of the Tree of Life' thus illustrates the genetic connections between all living things." http://tolweb.org/tree/.

The ***Marine Life Information Network*** is a great resource for Atlantic marine life information. From the site: "The Marine Life Information Network—MarLIN is an initiative of the Marine Biological Association of the UK (MBA). MarLIN pioneered the use of the Web for the dissemination of quality assured information on marine biodiversity of the North East Atlantic. In the last ten years, MarLIN has become the most comprehensive source of information on the marine biodiversity of the North East Atlantic. Our information is made freely and rapidly available through the Internet. The program has developed in collaboration with the major environmental protection agencies in the United Kingdom together with academic institutions." http://www.marlin.ac.uk.

Produced by the University of California–Berkeley, ***The Paleontology Portal*** features galleries of ancient life-forms from the region today called North America, as well as interviews with paleontologists and information about courses and field sites. http://www.paleoportal.org.

The University of Michigan's Museum of Zoology ***Animal Diversity Web*** is an online database of "animal natural history, distribution, classification, and conservation biology." It has great information on thousands of animal species. http://animaldiversity.ummz.umich.edu/site/index.html.

The US Department of Agriculture's ***Plants Database*** "provides standardized information about the vascular plants, mosses, liverworts, hornworts, and lichens of the U.S. and its territories." http://plants.usda.gov/.

Birdlife International's ***Data Zone*** allows you to search for basic information on over ten thousand kinds of birds, searching their database either by common or by scientific names. http://www.birdlife.org/datazone.

The internationally maintained ***World Register of Marine Species*** (***WoRMS***) has a database providing scientifically reviewed information for over two hundred thousand species of marine life, including over twelve thousand images (as of August 2010). Go to http://www.marinespecies.org/, click on "Search taxa" (*taxa* simply refers to kinds of life), and type in the life-form you're after. The many results that pop up include (if available) notes on habitat, distribution, and so on, as well as links to useful external sites.

The Public Broadcasting Service's ***Evolution*** pages, assembled by WGBH Boston and the NOVA Science Unit, provide excellent videos featuring interviews with prominent biologists as well as learning resources for both teachers and students. http://www.pbs.org/wgbh/evolution/.

The National Center for Science Education at http://ncse.com/about ("a not-for-profit, membership organization providing information and resources for schools, parents and concerned citizens working to keep evolution in public school science education") has an ***Evolution Education*** page leading to many resources for students, including online access to the journal *Creation/Evolution*. http://ncse.com/.

Science Daily reports on scientific discoveries at large, including evolution. For information on evolution in general, visit http://www.sciencedaily.com/news/plants_animals/evolution/.

The University of Wisconsin–Madison's ***Primate Information Network*** has useful fact sheets on many primate species. A particularly fun feature of the site is the "Callicam," a camera mounted in a marmoset enclosure (marmosets are primates of the genus *Callithrix*). Main site:

http://pin.primate.wisc.edu/. To view the Callicam (which you can control yourself), go to http://pin.primate.wisc.edu/callicam/ (best viewing times are between 7 a.m. and 5 p.m. CST.)

The Smithsonian Institution's ***National Zoological Park*** website has photo galleries of many animals, including primates. Go to http://nationalzoo.si.edu/ and click on "Photo Galleries."

INDEX

Pages numbers in **bold** indicate figures and tables

INDEX

INDEX

INDEX

INDEX

INDEX

INDEX

INDEX

INDEX

INDEX

INDEX